碳素养与低碳生活

CARBON LITERACY AND LOW CARBON LIFESTYLE

张　沁 / 编著

中国环境出版集团 · 北京

图书在版编目（CIP）数据

碳素养与低碳生活 = Carbon Literacy and Low Carbon Lifestyle : 汉文、英文 / 张沁编著. -- 北京 : 中国环境出版集团, 2024.1

ISBN 978-7-5111-5678-5

Ⅰ. ①碳… Ⅱ. ①张… Ⅲ. ①节能－基本知识－汉、英 Ⅳ. ①TK01

中国国家版本馆CIP数据核字(2023)第216510号

出 版 人　武德凯
责任编辑　曲　婷
文字编辑　苗慧盟
装帧设计　彭　杉

出版发行　中国环境出版集团
（100062 北京市东城区广渠门内大街16号）
网　　址：http://www.cesp.com.cn
电子邮箱：bjgl@cesp.com.cn
联系电话：010-67112765（编辑管理部）
发行热线：010-67125803 010-67113405（传真）
印　　刷　玖龙（天津）印刷有限公司
经　　销　各地新华书店
版　　次　2024年1月第1版
印　　次　2024年1月第1次印刷
开　　本　889×1194 1/16
印　　张　6
字　　数　328千字
定　　价　80.00元

编者的话

Editor's Message

PROF. KAREN CHEUNG
张沁教授

President of UNESCO Hong Kong Association
香港联合国教科文组织协会　会长

Director of Hong Kong Institue of Education for Sustainable Development
香港可持续发展教育学院　院长

After experiencing primitive, horticultural, and industrial civilizations, the mankind is now progressing into a new stage of ecological civilization. To practise President Xi Jinping's Thought on Ecological Civilizations, it is vital to incorporate the notions of carbon emissions peak and carbon neutrality into the overall ecological planning. Such ecological planning is part of the great mission of the era to achieve the rejuvenation of the Chinese nation. Inspired by Xi's Thought on Ecological Civilization, ecological education will be the predominant path to achieve the 2050 Vision of Beautiful China, and the United Nations Sustainable Development Goals.

In 2023, the United Nations made a global appeal for "Towards A Net Zero Future", where our nation has demonstrated her sense of duty as a great nation in leading sustainable development. At the United Nations General Debate in 2020, President Xi has announced a goal to reach carbon emissions peak by 2030 and carbon neutrality by 2060, which pioneers international development towards net zero.

In response to the international and national appeals, carbon literacy and low carbon lifestyle have become a compulsory element in Chinese civic education, which calls for collaborative efforts of cities, regions, organizations and individuals, so that we can implement carbon reduction targets.

The bilingual textbook "Carbon Literacy and Low Carbon Lifestyle" analyses both the causes and effects of climate change. It explicates the fundamental logic of climate action, including the components and reasons behind it. The textbook further explores the multiple factors behind climate change through science, in order to discover the real cause among all. By studying international and national policy frameworks, the textbook also responds to the real cases from top-level design to regional applications. Such doctrine illustrates the scenarios applying new solutions, inspiring critical thinking regarding causes and solutions. In the long run, it transforms knowledge into individual and collective climate actions.

As a bilingual teaching material, this textbook cites excerpts from various international academic journals and standards. It also introduces the conception, values and application of low carbon lifestyle by using plain language, graphical illustration and a variety of real-life examples. Acts such as energy saving, waste reduction, usage of renewable energies, participation in green action exemplify the value of low carbon lifestyle on both personal and society level, hence its significance in achieving sustainable development.

On the basis of UNESCO's vision of "Lifelong Learning", the application of this textbook is not merely limited to teaching in primary and secondary schools, colleges and universities. It can also be widely used

编者的话 Editor's Message

by to professionals of all ages and in various industries. The textbook adopts a highly-flexible approach, where actions for carbon neutrality can immerse into different aspects of life, including but not limited to the campus and workplace setting. In other words, this approach advocates for collaborative actions from the entire population, showcasing our approach to the global community.

With the fine vision of a sustainable future, this textbook aims to lead collaborative climate action from all students, teachers, corporations, families and communities. It focuses on the way to achieve a sustainable low carbon lifestyle, which accelerates the achievement of sustainable development and the harmonious coexistence between human and nature.

Credit and appreciation are given to Miss Americana Chen, the assistant editor and all editing staff involved from the Hong Kong Institute of Education for Sustainable Development and the publishing house for the successful publication of the book.

人类社会在经历了原始文明、农业文明、工业文明之后，目前已进入了崭新的生态文明发展时代。深入学习贯彻习近平生态文明思想，“把碳达峰、碳中和纳入生态文明建设整体布局”，是实现中华民族伟大复兴的历史使命。在习近平生态文明思想指引下，生态教育将会愈加成为实现 2050 年美丽中国愿景、促进联合国可持续发展目标实现的重要途径。

2023 年，联合国发出了“迈向净零排放未来”的全球呼吁，在联合国推进可持续发展目标中，中国展现负责任的大国担当，以习近平主席于 2020 年在联合国大会一般性辩论上向全世界宣布的“中国将力争 2030 年前实现碳达峰、2060 年前实现碳中和”为目标，引领全球迈向净零排放的步伐。

为响应国际及国家号召，提高碳素养、践行低碳生活已成为每位公民的必修内容，只有凝聚每个城市、地区、机构、个人的力量才能够更高效地将减碳目标贯彻落实。

《碳素养与低碳生活》双语教材从气候变化的成因与影响着手，清晰表达出气候行动的底层逻辑，解答了气候行动“是什么？为什么？”，进而从科学中找出多元因素，发现问题，更配合国际、国家政策框架，从顶层设计到各地区项目的实践案例，清楚展现各实用场景，鼓励学习者勇于思考、积极探寻问题解决方案，将知识转化为个人及集体应对气候变化的实际行动。

作为双语教材，本书中引用了诸多国际文件及标准，尽可能引述官方中英文原文，辅以深入浅出的语言、清晰的图片和丰富的实例介绍低碳生活的概念、价值和实践方法（包括通过节能、减少浪费、使用可再生能源、参与环保行动等方式来减少对环境的影响），展现出低碳生活的个人价值和社会价值，以及对实现可持续发展的重要意义。

在联合国教科文组织“终身学习”的主张下，本书不仅适合中小学、高等院校在学校课堂、校园内外活动中使用，更可广泛用于各行业、各年龄的专业人士。本书灵活性高，可融入校园、职场、生活等各个场景下的碳中和行动，协助推动全面低碳行动，向全球展现我们“怎么做”。

怀揣着对可持续未来的美好愿景，本书致力于带领每一位师生、每一所学校、每一个机构、每一个家庭、每一个社区，携手开展低碳行动，聚焦可持续低碳生活，加速实现可持续发展及人与自然的和谐共生。

感谢香港可持续发展教育学院助理编辑陈思桦以及出版社的各位编辑为本书的顺利出版所做的努力。

目录 Table of Contents

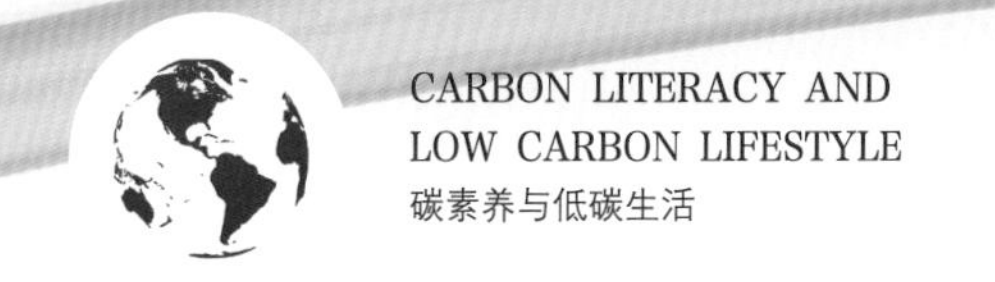

目录 Table of Contents

CLIMATE CHANGE SCIENCE

气候变化的科学

What is Climate Change?
什么是气候变化?

Weather 天气	V/S	Climate 气候
Atmosphreic condition at a specific point of time 任何在特定时间的大气状况		Average weather conditions over a longer period of time(e.g.30 years) 较长时间维度内的平均天气状况（如 30 年）

Climate in a narrow sense is usually defined as the average weather, or more rigorously, as the statistical description in terms of the mean and variability of relevant quantities over a period of time ranging from months to thousands or millions of years. The classical period for averaging these variables is 30 years, as defined by the World Meteorological Organization. The relevant quantities are most often surface variables such as temperature, precipitation and wind. Climate in a wider sense is the state, including a statistical description, of the climate system. (IPCC, 2013)

狭义上，气候通常被定义为“天气的平均状况”，或更严格地表述为在某一个时期内对相关变量的平均值和变化率作出的统计描述。这个时期的长度从几个月至几千年甚至几百万年不等，而世界气象组织 (WMO) 将其定义为 30 年。这些相关量一般指地表变量，如温度、降水和风。广义上，气候就是气候系统的状态，包括统计上的描述。（联合国政府间气候变化专门委员会，2013）

Scientists are asked how they can accurately predict climate 50 years from now when they cannot predict the accurately weather a few weeks from now. The chaotic nature of weather makes it unpredictable beyond a few days. Projecting changes in climate (i.e., long-term average weather) due to changes in atmospheric composition or other factors is very different and much more manageable.

As an analogy, while it is impossible to predict the age at which any particular man will die, we can say with high confidence that the average age of death for men in industrialized countries is about 75.

科学家们常被问道：如果连数周之后的天气都无法准确预测，又如何准确预测 50 年后的气候呢?

这是因为天气的混沌性质令我们难以准确预测数天后的天气。然而，根据大气成分或其他因素来预估气候的变化（也就是长期的平均天气）是非常不同且相对较易解决的问题。譬如，我们不可能准确预测某人的死亡年龄，但我们可以很有把握地说，工业国家的居民的平均死亡年龄大约是 75 岁。

Simply speaking, climate tells you what clothes to buy, weather tells you what clothes to wear.

简单来说，气候告诉你该买什么衣服，天气告诉你该穿什么衣服。

Which of the following are "climate"?（See next page for answer）
以下哪些是“气候”？（答案见下页）

The dry season in Kenya is from June to October. 7—10 月是肯尼亚的旱季。	Freezing temperatures in Toronto for 7 days in a row. 多伦多出现连续 7 天极度低温。	Today's temperature in Beijing feels hotter than usual. 今天感觉北京比平时更热。	Thailand has high humidity all year round. 泰国全年的空气湿度都很高。

Source 资料来源：IPCC 联合国政府间气候变化专门委员会。

What is Climate Change?
什么是气候变化?

Complexity of the Global Climate System
全球气候系统的复杂性

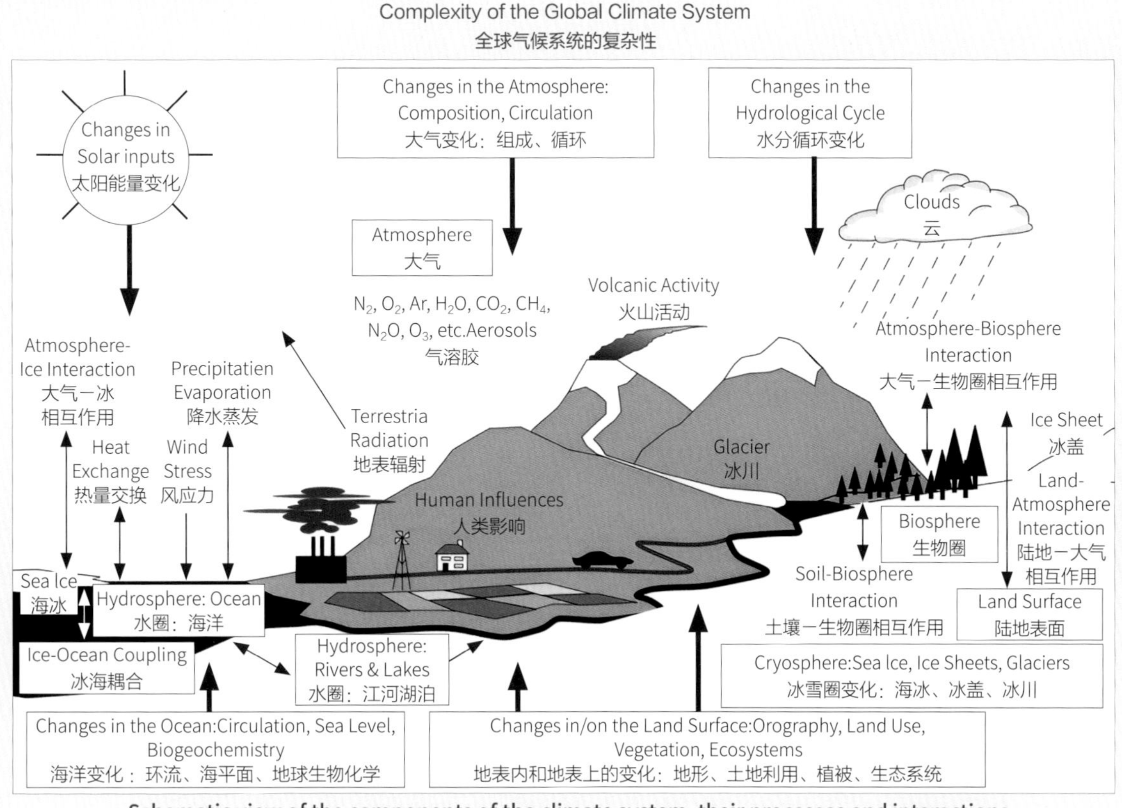

Schematic view of the components of the climate system, their processes and interactions
气候系统各组成部分、其过程和相互影响

Source 资料来源： IPCC 联合国政府间气候变化专门委员会。

Climate in a wider sense is the state, including a statistical description, of the climate system, which consists of five major components.
广义上，气候是气候系统的状态，包括统计描述，有 5 个主要组成部分。(世界气象组织)

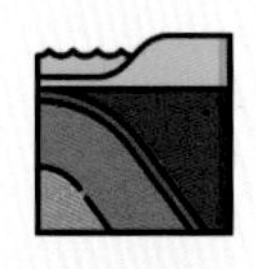
Lithosphere
岩石圈

Atmosphere
大气层

Cryosphere
冰雪圈

Hydrosphere
水圈

Biosphere
生物圈

Climate change refers to a change in the state of the climate that can be identified by changes in the mean and/or the variability of its properties and that persists for an extended period, typically decades or longer. Including temperature, humidity, precipitation and extreme weather.

气候变化是指气候状态的变化，可以通过平均值的变化和/或其属性的变化来识别，并且持续很长一段时间(通常是几十年或更长时间)，包括温度、湿度、降水和极端天气。

Answer Key 答案(P.2)： Climate, Weather, Weather, Climate
气候、天气、天气、气候

What is Climate Change?
什么是气候变化?

What causes change in climate?/ 什么会导致气候的变化?

Earth's climate does not change without a reason. Many factors may influence it over long periods of time. Such factors are known as "climate forcing".

There are 3 main climate drivers:

地球的气候不会无缘无故地改变。许多因素都可能对气候产生长远影响，这些因素被称为“气候强迫因子”。其中包括 3 个主要的气候诱因：

Solar Variability
太阳变率

Solar variability refers to the changes in the levels of solar radiation with 11-year cyclic variation. However, most of the current research and observations suggest that global warming is overtaking the solar role.

太阳变率是指太阳辐射量在 11 年的周期内的变化。然而，目前的大多数研究和观察表明，全球变暖的影响远大于太阳的作用。

Source 资料来源：Kumar P, et al.。

Volcanic Activities
火山爆发

Volcanic eruptions can also cause climate disruption. Large eruptions inject aerosols into the atmosphere, which cools the Earth's surface. For example, an eruption such as that of Mt. Pinatubo in 1991, reduced the average global temperature of about 0.6°C over the next 15 months.

火山爆发也会影响气候。大型的火山爆发会将气溶胶注入大气，令地球表面变冷。例如，1991 年皮纳图博火山爆发使全球平均温度在 15 个月中降低了 0.6℃。

Source 资料来源：NASA Goddard Space Flight Center 戈达德太空飞行中心。

Change in Carbon Cycle
碳循环的改变

Oxygen manufactured by plants and consumed by animals, and carbon dioxide, which is released by animals and used by plants has a give and take relationship. When the amount of carbon dioxide and other greenhouse gases in the atmosphere increases, this has a direct causal effect on changes in the climate.

动物会消耗植物制造的氧气，而植物则靠动物释放的二氧化碳生存，两者互相依存。当大气中的二氧化碳和其他温室气体的含量增加时，会对气候变化产生直接影响。

Source 资料来源：US EPA 美国国家环境保护局。

Discussion
讨论议题

How does the climate forcing affects different components of the Earth's Climate System?
这些气候强迫因子如何影响地球气候系统中不同的构成部分?

Which climate forcing has had the greatest impact in recent years?
哪种气候强迫因子在近年来对地球气候系统造成最大的影响?

What is Climate Change?
什么是气候变化？

Global Warming/ 全球变暖

Since the early 20th century, scientists have observed changes in climate that cannot be explained by any natural climate change in the past. This is global warming, the fastest climate change on record (WMO).

According to the United Nations, greenhouse gas concentrations are at their highest levels in 2 million years. As a result, the Earth is now about 1.1°C warmer than it was in the late 1800s. The last decade (2011—2020) was the warmest on record.

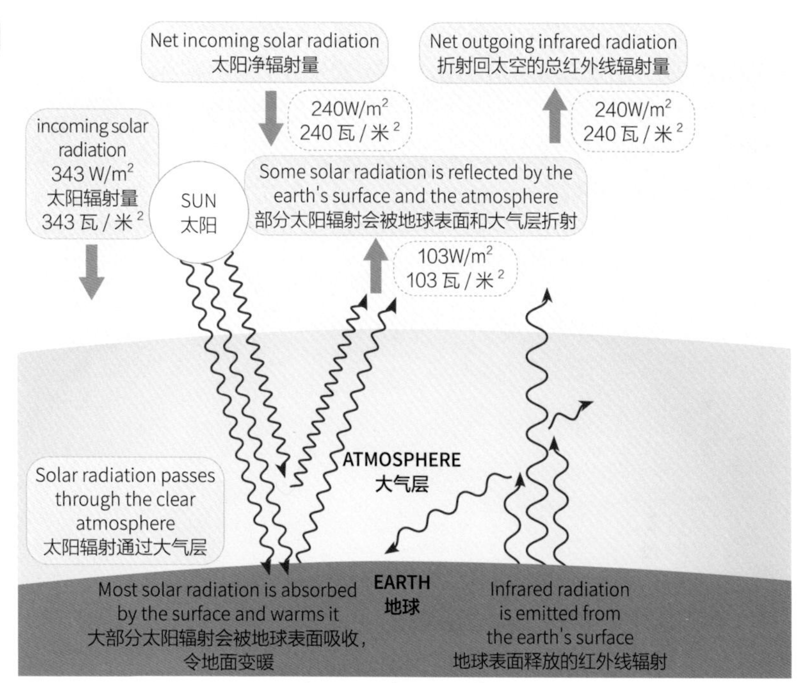

The Greenhouse Effect
温室效应

20 世纪初，科学家观测到了以往任何自然气候变化都无法解释的现象，这正是有史以来发生得最快的气候变化——全球变暖（世界气象组织）。联合国指出，地球的温室气体浓度正处于 200 万年来的最高水平。截至目前，地球的温度较 19 世纪末升高了约 1.1℃。过去 10 年（2011—2020 年）的温度达到历史之最。

The Greenhouse Effect
温室效应

The Greenhouse Effect refers to a physical property of the Earth's atmosphere. If the Earth had no atmosphere, its average surface temperature would be very low at about-18°C rather than the comfortable 15°C found today. The difference in temperature is due to a suite of gases called greenhouse gases which affect the overall energy balance of the Earth's system by absorbing infra-red radiation. In its existing state, the Earth-atmosphere system balances absorption of solar radiation by emission of infrared radiation to space. Due to greenhouse gases, the atmosphere absorbs more infrared energy than it re-radiates to space, resulting in a net warming of the Earth-atmosphere system and of surface temperature. This is the Natural Greenhouse Effect. With more greenhouse gases released to the atmosphere due to human activities, more infrared radiation will be trapped in the Earth's surface which contributes to the Enhanced Greenhouse Effect.

温室效应是指发生在地球大气层的一种物理现象。若没有大气层，地球表面的平均温度将会降低至 -18℃，而非现在合宜的 15℃。这种温度上的差异由名为“温室气体”的一类气体引起，它们通过吸收红外线辐射，影响地球整体的能量平衡。一般而言，地面和大气层会将红外线辐射释放到太空，维持整体平衡以吸收太阳幅射。受到温室气体的影响，大气层吸收红外线辐射的量比释放到太空的多，令地球表面温度上升，这便是“天然温室效应”。由于人类活动也释放出大量温室气体，更多红外线辐射被困在地球表面上从而导致现在的温室效应增强。

Source 资料来源：NRDC 自然资源保护协会，Hong Kong Observatory 香港天文台。

Anthropogenic Causes of Climate Change
造成气候变化的人为因素

Climate change and global warming are amongst the biggest current challenges of humankind. Since the 1800s, human activities have been the main driver of climate change, primarily due to the burning of fossil fuels like coal, oil and gas, which generates greenhouse gas emissions and warms up our planet.

气候变化和全球变暖是人类当前面临的最大挑战之一。自 19 世纪以来，人类活动一直是影响气候变化的主要原因，特别是燃烧煤炭、石油和天然气等化石燃料所产生的温室气体排放，使地球温度不断升高。

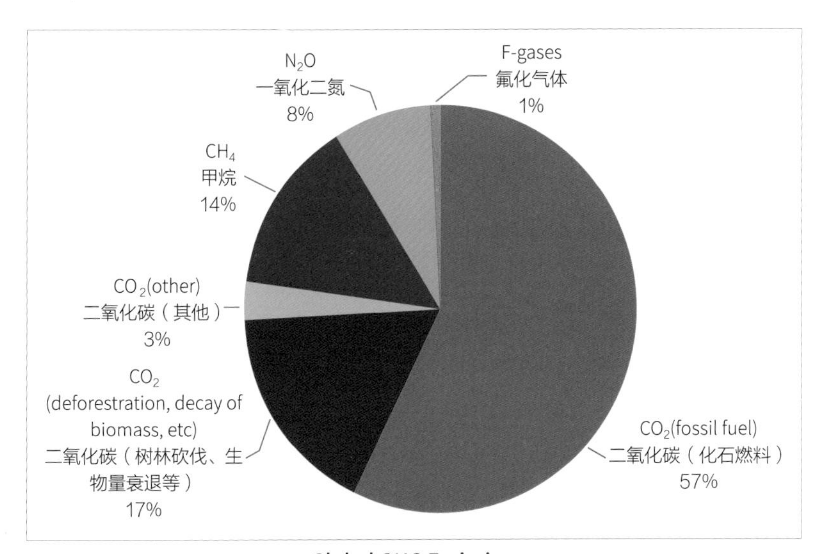

Global GHG Emission
全球温室气体排放情况

Source 资料来源：EMSD HK 香港机电工程署。

The main Greenhouse Gases (GHG) include:

- Carbon Dioxide CO_2
- Methane CH_4
- Nitrous Oxide N_2O
- Hydrofluoro-carbons
- Perfluoro-carbons
- Sulphur Hexafluoride

主要温室气体包括：

- 二氧化碳 CO_2
- 甲烷 CH_4
- 一氧化二氮 N_2O
- 氢氟烃 HFCs
- 全氟化碳 PFCs
- 六氟化硫 SF_6

What are Greenhouse Gases?
温室气体是什么？

Greenhouse gases trap heat in the atmosphere and warm the planet as they enhance the Greenhouse effect. Greenhouse gases have different chemical properties and are removed from the atmosphere, over time, by different processes.

The level of influence of greenhouse gases on global warming depends on three key factors:

- **Concentrations:** measured in parts per million (PPM),1 PPM for a given gas means that there is 1 molecule of that gas in every 1 million molecules of air.
- **Lifetime:** how long it remains in the atmosphere.
- **Global Warming Potential (GWP):** how effective it is at trapping heat, a measure of the total energy that a gas absorbs over a given period of time (usually 100 years) relative to the emissions of 1 ton of carbon dioxide.

温室气体会在大气中捕获热量并增强温室效应，使地球变暖。温室气体具有不同的化学性质，其随着时间通过不同方式从大气中消除。

温室气体对全球变暖的影响程度取决于 3 个关键因素：

- **浓度：**以百万分率 (PPM)* 为单位测量，给定气体的 1PPM 意味着每 100 万个空气分子中就有一个该气体分子。
- **寿命：**气体在大气中停留的时间。
- **全球变暖潜能 (GWP)：**它捕获热量的能力，衡量一种气体在给定时间段（通常为 100 年）内吸收的总能量（相对于 1 吨二氧化碳排放量）。

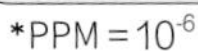

*PPM = 10^{-6}
Source 资料来源：NRDC 自然资源保护协会。

Anthropogenic Causes of Climate Change
造成气候变化的人为因素

CarbonDioxide
二氧化碳 CO_2

What is carbon dioxide?

- The most important Greenhouse Gas
- Traps heat in the atmosphere
- the basis for comparing the global warming potential of other greenhouse gases, emissions of other gases can be converted into carbon dioxide equivalents (CO_2e)

Sources:

- burning fossil fuels (coal, natural gas, and oil), solid waste, trees and other biological materials
- A result of certain chemical reactions (e.g., manufacture of cement)

How is CO_2 removed from the atmosphere?

- Absorbed by plants via photosynthesis
- Ocean Uptake

Importance for climate:

- Absorbs infrared radiation
- affects stratospheric Ozone (O_3)

什么是二氧化碳?

- 最重要的温室气体
- 在大气中捕获热量
- 作为比较其他温室气体全球变暖潜能的基础，其他气体的排放可以换算成二氧化碳当量

来源:

- 化石燃料（煤、天然气和石油）燃烧、固体废物、树木和其他生物材料
- 某些化学反应的结果（如水泥制造）

如何从大气中去除二氧化碳?

- 通过植物光合作用吸收
- 被海洋吸收

对气候的影响:

- 吸收红外线辐射
- 影响大气平流层中臭氧（O_3）的浓度

Scripps Institution of Oceanography
NOAA Global Monitoring Loborotory
斯克里普斯海洋研究所
NOAA 全球监测实验室

Carbon dioxide concentration /10PPM
二氧化碳浓度 /10PPM
420 400 380 360 340 320
1960 1970 1980 1990 2000 2010 2020
Year/ 年份

Atmospheric CO_2 at Mauna Loa Observatory
大气层中的二氧化碳

Source 图表来源：U.S.NOAA 美国国家海洋大气局。

From 2000 to 2021, the global atmospheric CO_2 amount has grown by 43.5 PPM, an increase of 12%!

从 2000—2021 年，全球大气层二氧化碳浓度上升了 43.5 PPM，相较于 2000 年增长了 12%！

Source 资料来源：National Oceanic & Atmospheric Administration(NOAA) 美国国家海洋大气局。

Another reason carbon dioxide is important in the Earth system is that it dissolves into the ocean like the fizz in a can of soda, which lowers the ocean's pH (raising its acidity). Since the start of the Industrial Revolution, the pH of the ocean's surface waters has dropped from 8.21 to 8.10. This drop in pH is called ocean acidification.

A drop of 0.1 may not seem like a lot, but the pH scale is logarithmic; a 1-unit drop in pH means a tenfold increase in acidity. A change of 0.1 means a roughly 30% increase in acidity. Increasing acidity interferes with the ability of marine life to extract calcium from the water to build their shells and skeletons.

另外，二氧化碳对地球系统很重要的原因是它会像汽水一样在海洋中溶解，从而降低海洋的酸碱度（降低 pH 提高其酸度）。自工业革命以来，海洋地表水的 pH 已从 8.21 降至 8.10，这种现象被称为“海洋酸化”。

下降 0.1 可能看起来不多，但从 pH 的衡量比例来看，pH 下降 1 个单位意味着酸度增加 10 倍。0.1 的变化意味着酸度增加了大约 30%。酸度增加会影响海洋生物从水中提取钙来生成贝壳和骨骼的能力。

Anthropogenic Causes of Climate Change
造成气候变化的人为因素

Methane
甲烷 CH_4

Sources:
- Biomass burning
- Enteric fermentation
- Rice paddies

来源:
- 生物体的燃烧
- 肠道发酵作用
- 水稻

How is CH_4 removed from the atmosphere?
- Reactions with Hydroixide (OH)
- Microorganisms uptake by soils

如何从大气中去除甲烷?
- 和氢氧化物（OH）发生化学作用
- 被土壤内的微生物吸收

Importance for climate:
- Absorbs infrared radiation
- affects tropospheric O_3 and OH
- affects stratospheric O_3 and water vapour (H_2O)
- produces CO_2
- relatively short life cycle

对气候的影响:
- 吸收红外线辐射
- 影响对流层中的 O_3 及 OH 的浓度
- 影响平流层中 O_3 和水汽（H_2O）的浓度
- 产生 CO_2
- 生命周期较短

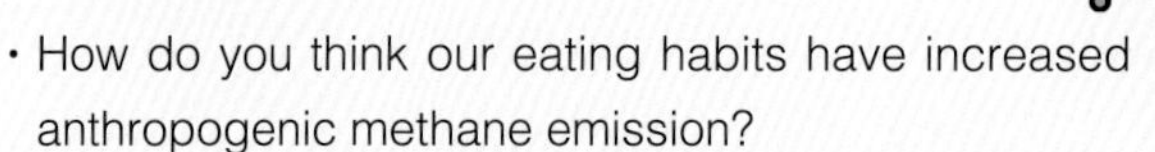

Discussion
讨论议题

- How do you think our eating habits have increased anthropogenic methane emission?
- 你认为我们的饮食习惯如何加剧了人为甲烷排放?
- Could you suggest some ways that we can adopt more sustainable diet habits?
- 你能否提出一些我们可以转向更可持续的饮食习惯的方法?

Nitrous Oxide
一氧化二氮 N_2O

Sources:
- Agriculture
- Industry
- Fossil-fuel combustion
- Biomass burning
- Waste water treatment

来源:
- 农业
- 工业
- 化石燃料燃烧
- 生物质燃烧
- 废水处理

How is N_2O removed from the atmosphere?
- Removal by soils
- Stratospheric photolysis and reaction with Oxygen (O)

如何从大气中去除一氧化二氮?
- 被土壤吸收
- 在大气平流层中被光解及与氧原子 (O) 产生化学作用

Importance for climate:
- Absorbs infrared radiation
- Affects stratospheric O_3

对气候的影响:
- 吸收红外线辐射
- 影响大气平流层中 O_3 的浓度

Source 资料来源：National Oceanic & Atmospheric Administration (NOAA) 美国国家海洋大气局。

Anthropogenic Causes of Climate Change
造成气候变化的人为因素

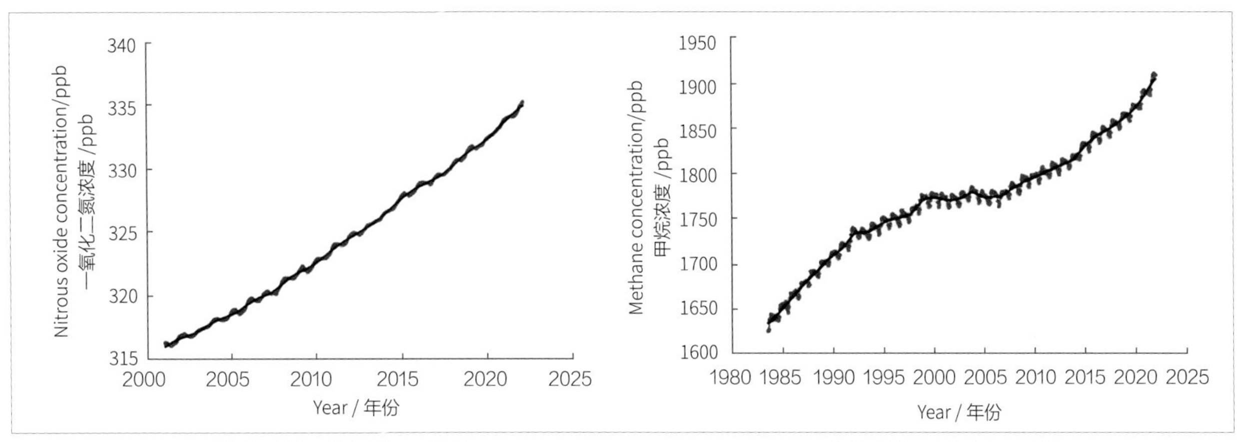

Global Monthly Mean N_2O
全球一氧化二氮每月平均值

Global Monthly Mean CH_4
全球甲烷每月平均值

Source 图表来源：U.S.NOAA 美国国家海洋大气局。

Fill in the correct numbers.（See next page for answers）
填入正确的数字。（答案见下页）

(See next page for answers)

Agriculture, Forestry and Other Land Use (AFOLU) activities accounted for around ________ of CO_2, ________ of methane (CH_4), and ________ of nitrous oxide (N_2O) emissions from human activities globally during 2007—2016, representing ________ of total net anthropogenic emissions of GHGs! (IPCC, 2019, Special Report on Climate Change and Land)

2007—2016 年，农业、林业和其他土地利用 (AFOLU) 活动所产生的各种温室气体分别占全球人类活动产生的二氧化碳排放量的 ________ 左右、甲烷 (CH_4) 排放量的 ________ 左右和一氧化二氮 (N_2O) 排放量的 ________ 左右，约占人为排放的温室气体总量的 ________！（联合国政府间气候变化专门委员会，2019，特别报告：气候变迁和土地报告）

Fluorinated Gases(F-gases)
氟化气体

Including Hydrofluoro-carbons (HFCs), perfluorocarbons (PFCs) and sulfur hexafluoride (SF_6)

Sources:

- Refrigerants and air conditioners
- Industrial processes such as aluminum and semiconductor manufacturing

How is F-gases are removed from the atmosphere?

- Destroyed by sunlight in the far upper atmosphere

Importance for climate:

- High Global Warming Potential (GWP)
- Long lifetime

包括氢氟烃 (HFCs)、全氟化碳 (PFCs) 和六氟化硫 (SF_6)

来源：

- 制冷剂、空调设备
- 铝和半导体制造等工业过程

如何从大气中去除氟化气体？

- 在遥远的高层大气中被阳光摧毁

对气候的影响：

- 高全球变暖潜能
- 在大气中留存的时间长

Source 资料来源：National Oceanic & Atmospheric Administration (NOAA) 美国国家海洋大气局。
ppb = 10^{-9}

Anthropogenic Causes of Climate Change
造成气候变化的人为因素

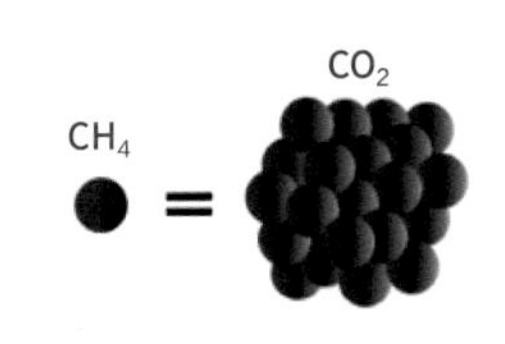

Global Warming Potential
全球变暖潜能

Gas 气体	Sources 来源	Global Warming Potential (GWP) (100 years) 全球变暖潜能（100 年）
Carbon Dioxide 二氧化碳 CO_2	Fuel & Electricity Consumption 燃料和电力消耗	1
Methane 甲烷 CH_4	Coal mines, Gas leakage, Landfill, Agriculture 煤矿、煤气泄漏、垃圾堆填区、农业	27.9
Nitrous Oxide 一氧二化氮 N_2O	Agricultural Soil Management & Mobile Source Fossil Fuel 农业土壤管理和移动源化石燃料	273
HFC - 134a 氢氟烃 - 134a	Refrigerant, Plastic Blowing Agent, Cleaning Solvent, Propellant 制冷剂、塑料发泡剂、清洗溶剂、推进剂	1 530
PFC-14 全氟化碳 -14	Semiconductor Manufacturing 半导体制造	7 380
Sulfur Hexafluoride 六氟化硫 SF_6	Electrical Transmission & Distribution Systems, Semiconductor Manufacturing 输配电系统、半导体制造	25 200

Source 资料来源：IPCC 联合国政府间气候变化专门委员会。

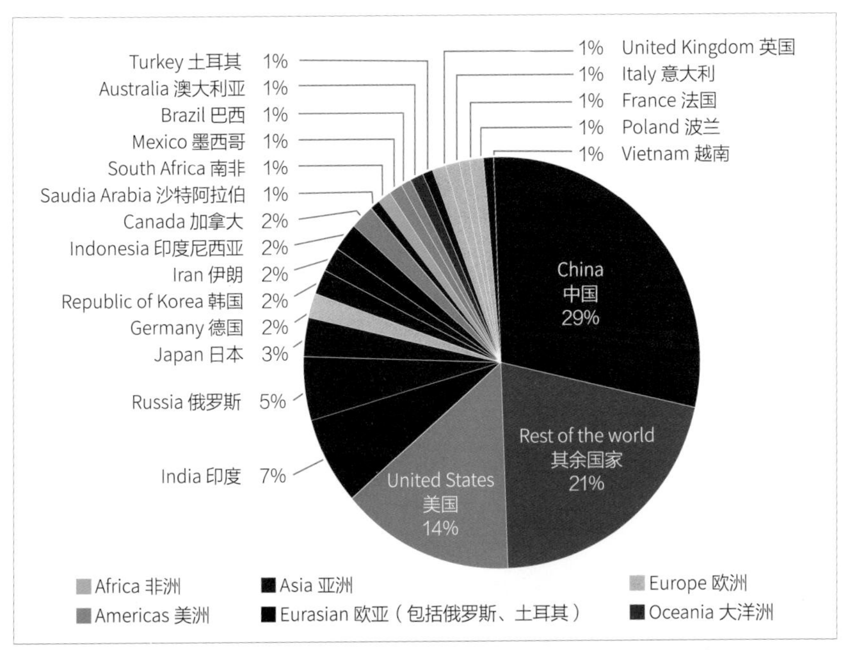

Top Annual CO_2 Emitting Countries, 2019 (from fossil fuels)
2019 年化石燃料的二氧化碳排放量最高的国家

Source 资料来源：Union of Concerned Scientists, 忧思科学家联盟；IEA Atlas of Energy 国际能源署。

Global Emission Share
全球排放量占比

In 2019, China emitted the greatest amount of CO_2 from the burning of fossil fuels, contributing to 29% of the worlds' total CO_2 emission from fossil fuels. The United States ranked the second by taking 14% of the total emission and India ranked the third with 7%.

2019 年，中国的化石燃料燃烧产生的二氧化碳排放量居全球首位，占全球化石燃料二氧化碳排放总量的 29%。第二为占比 14% 的美国，第三为占比 7% 的印度。

Answer Key 答案（P.9）： 13%，44%，81%，23%

Climate Change Trends
气候变化趋势

Since 1850, each of the last four decades has been successively warmer than any decade that preceded it. Global average surface temperature in the first two decades of the 21st century (2001—2020) was 0.99°C higher than 1850—1900. Global average surface temperature was 1.09°C higher in 2011—2020 than 1850—1900, with larger increases over land (1.59°C) than over the ocean (0.88°C).

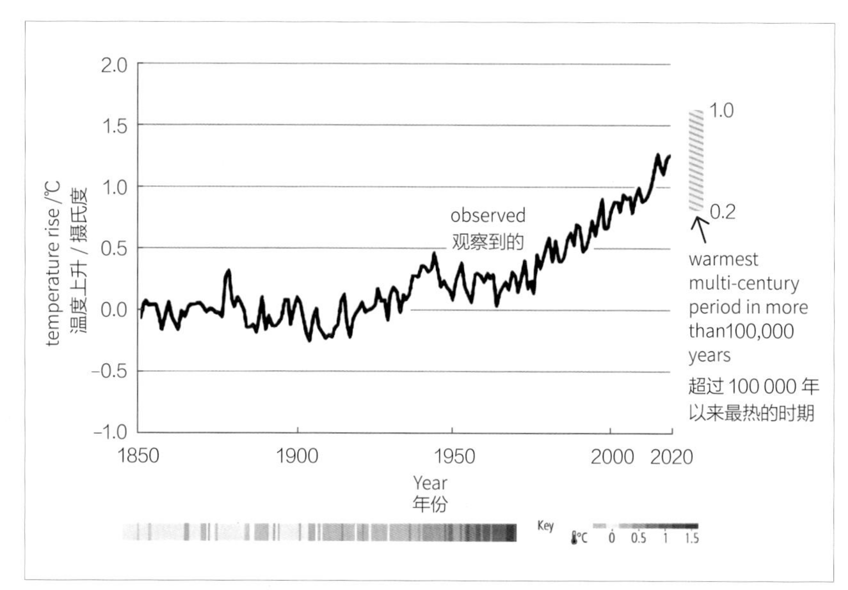

Global surface temperature has increased by 1.1℃ by 2011—2020 compared to 1850—1900

与 1850—1900 年相比，2011—2020 年全球地表平均温度增加了 1.09℃

Source 资料来源：IPCC, 2023,《Synthesis Report of the IPCC Sixth Assessment Report (AR6)》联合国政府间气候变化专门委员会，2023，《第六次评估报告的综合报告》。

自 1850 年以来，每一个 10 年期间的平均地表温度都创下历史新高。21 世纪前 20 年（2001—2020 年）的全球平均地表温度较 1850—1900 年高 0.99℃。2011—2020 年全球平均地表温度较 1850—1900 年高 1.09℃，其中陆地（1.59℃）的升幅大于海洋（0.88℃）。

The total human-caused global surface temperature increase from 1850–1900 to 2010–2019 is best estimated to be 1.07°C. It is likely that well-mixed GHGs contributed a warming of 1.0°C to 2.0°C. GHGs were the main driver of tropospheric warming since 1979.

从 1850—1900 年到 2010—2019 年，人为造成全球平均地表温度升高了 1.07℃。均匀混合的温室气体导致的气温升高高达 1.0℃至 2.0℃。自 1979 年以来，温室气体是导致对流层变暖的主要驱动因素。

Source 资料来源：IPCC 联合国政府间气候变化专门委员会。

Try to match the following sentences .（See next page for answer）

尝试配对以下句子。（答案见下页）

A. In 2017, CO_2 concentration in the atmosphere surpassed
2017 年，大气中 CO_2 浓度超过

B. Greenhouse gases remain in the atmosphere for years and
温室气体在大气中存留多年并且

C. In 2018, global annual CO_2 emissions surpassed
2018 年，全球二氧化碳年排放量超过

1. 405 PPM

2. 41.5 $GtCO_2$/year（年）

3. trap some of the energy that comes from the sun
捕获部分来自太阳的能量

Climate Change Trends
气候变化趋势

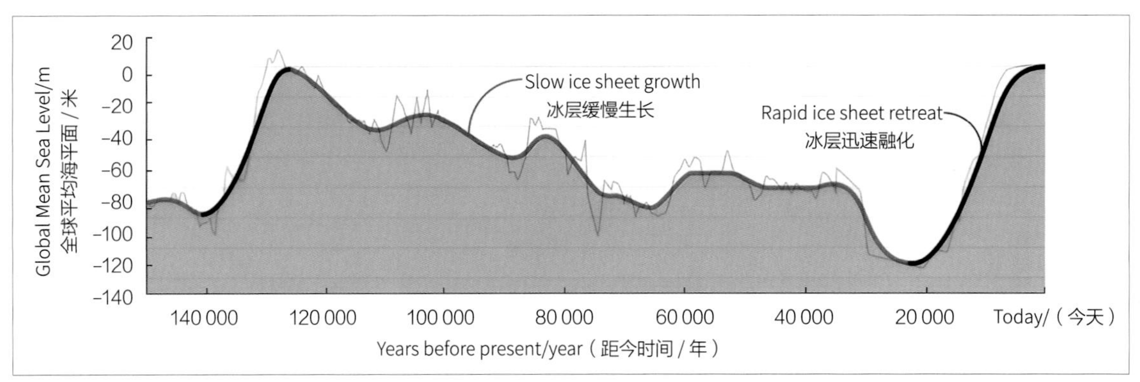

Change in Global Mean Sea Level/ 全球平均海平面变化

Across the globe, sea level is rising, and the rate of increase has accelerated. Sea level increased by about 4 mm per year from 2006 to 2018, which was more than double the average rate over the 20th century. Rise during the early 1900s was due to natural factors, such as glaciers catching up to warming that occurred in the Northern Hemisphere during the 1800s. However, since 1970, human activities have been the dominant cause of global average sea level rise.

Sea level rises either through warming of ocean waters or the addition of water from melting ice and bodies of water on land. Expansion due to warming caused about 50% of the rise observed from 1971 to 2018. Melting glaciers contributed about 22% over the same period. Melting of the two large ice sheets in Greenland and Antarctica has contributed about 13% and 7%, respectively, during 1971 to 2018, but melting has accelerated in the recent decades, increasing their contribution to 22% and 14% since 2016. Another source is changes in land-water storage: reservoirs and aquifers on land have reduced, which contributed about an 8% increasein sea level.

全球海平面正在上升，且上升的速度在加快。2006—2018 年，海平面每年上升约 4 毫米，是 20 世纪平均上升速度的两倍多。1900 年代初期，海平面上升主要是由自然因素造成，例如 1800 年代北半球的气候变暖令冰川融化。然而自 1970 年以来，人类活动一直是全球平均海平面上升的主要原因。

海平面上升的原因除了湖水的热力膨胀外，还有冰川融水以及陆地水体的汇入。1971—2018 年观测到的海平面上升中，约 50% 的扩张是由变暖造成。同期冰川融化则占约 22%。1971—2018 年，格陵兰和南极洲的两个大冰层融化分别约占 13% 和 7%，随着近几十年来融化速度加快，自 2016 年以来其比率分别增加至 22% 和 14%。另一个海平面上升的原因是陆地蓄水量的变化，由于陆地上的水库和含水层减少，导致海平面上升约 8%。

Can the melting of the ice sheets be reversed?
冰层的融化可以逆转吗？

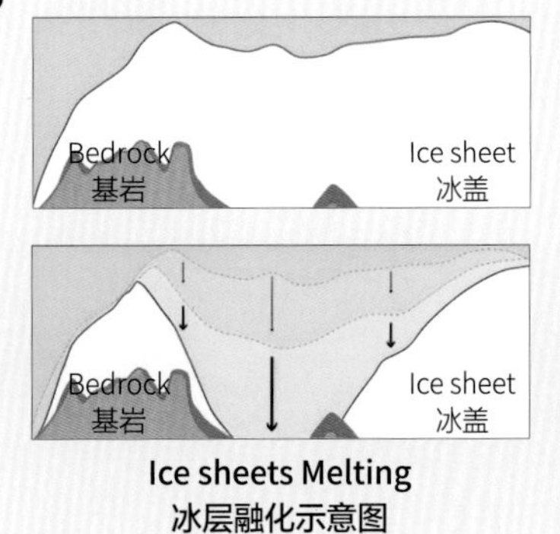

Ice sheets Melting
冰层融化示意图

Once ice sheets are destabilised, it takes them tens of thousands of years to re-grow.

冰层一旦变得不稳定，需要数万年才能重新形成。

In this picture, the ice sheet is very thick therefore its surface is very high and the air at high altitude is very cold.

图中的冰层很厚，因此它的表面很高，高海拔的空气很冷。

As the ice sheet melts, its surface goes down until it reaches a threshold, where the surrounding air is warmer and melts the ice even more quickly.

随着冰层融化，其表面的高度不断下降直到达到临界点，此时周围的空气温度更高，因此冰融化得更快。

Source 资料来源：IPCC 联合国政府间气候变化专门委员会。

Answer Key 答案（P.11）： A-1　B-3　C-2

Predictions of Future Climate Change
气候变化的预测

In 2014, the Intergovernmental Panel on Climate Change (IPCC) explores four potential futures depending on what policies governments adopt to cut emissions.

2014年，联合国政府间气候变化专门委员会(IPCC)根据政府采取的减排政策预测了4种可能发生的未来情景。

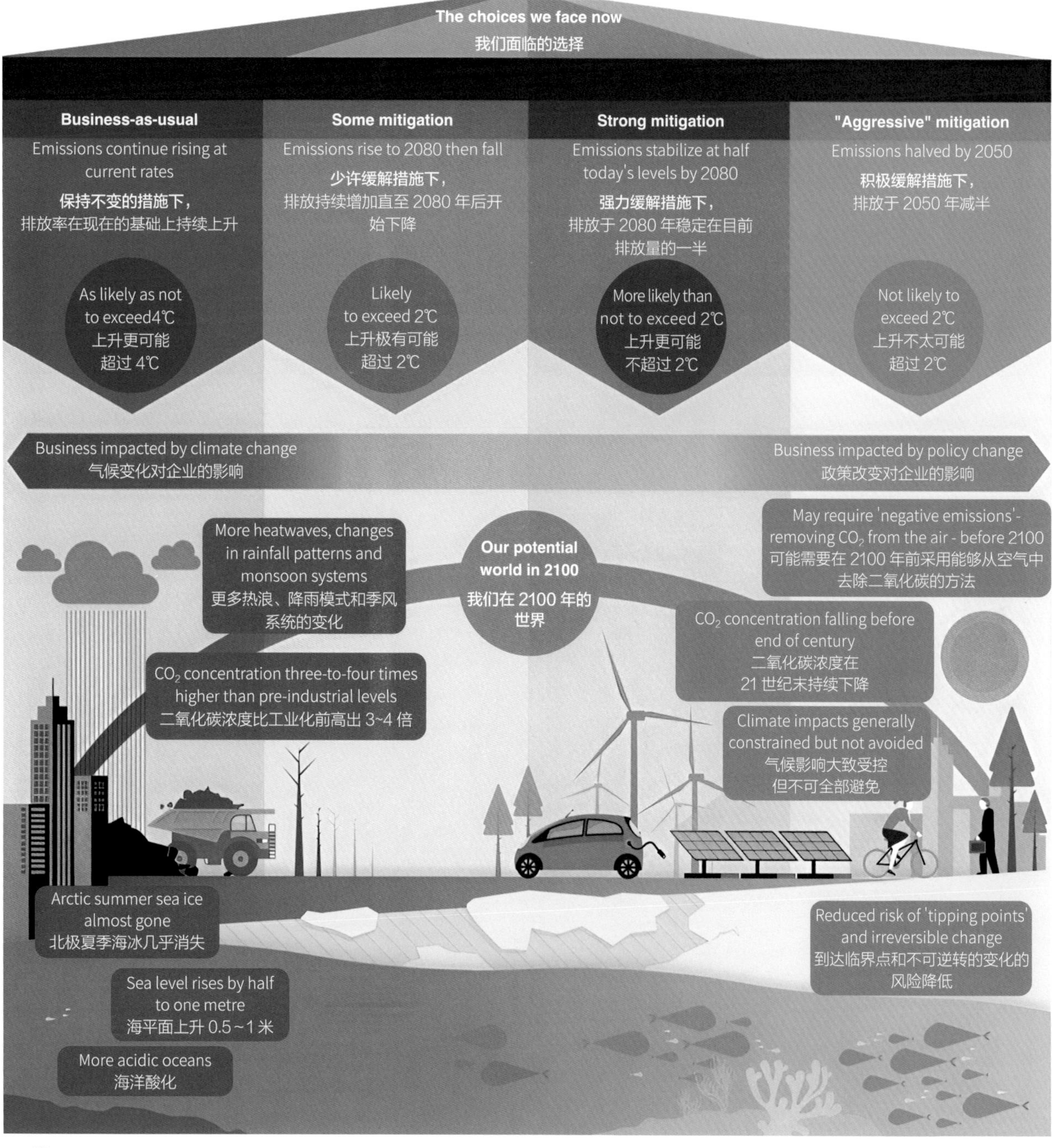

Discussion: Where do you think we're heading towards?
讨论议题：你认为我们在向哪个方向发展？

Source 资料来源：University of Cambridge Institute for Sustainability Leadership 剑桥大学可持续发展领导力研究所。

Predictions of Future Climate Change
气候变化的预测

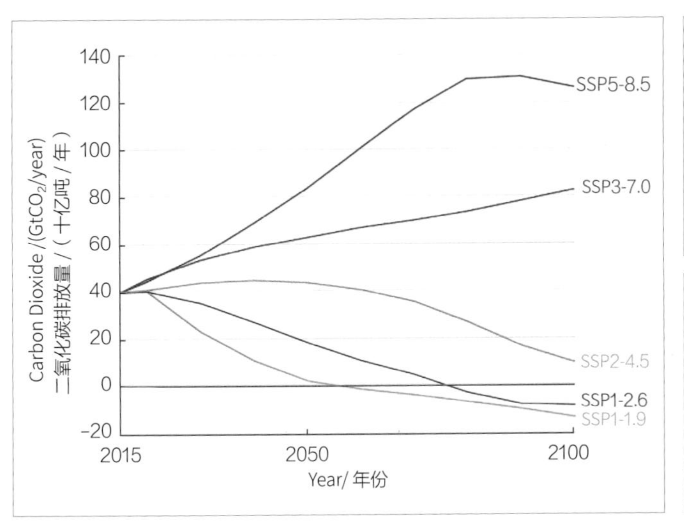

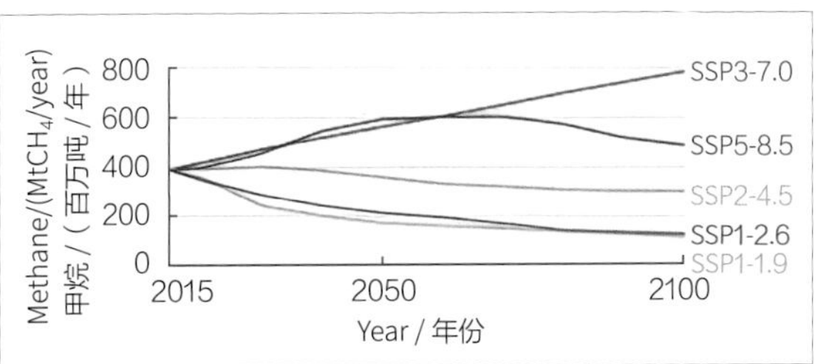

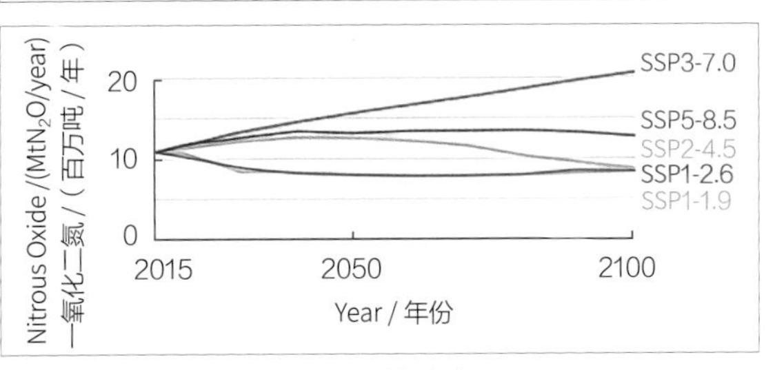

Future Annual Emissions of CO_2 (left) and of a subset of key non-CO_2 drivers (right), across five illustrative scenarios
在五个假设情景中，未来每年的二氧化碳排放量（左）和部分关键的非二氧化碳温室气体（右）排放量

The IPCC Report Climate Change 2021: The Physical Science Basis assesses the climate response to five illustrative scenarios of possible future development. They start in 2015, and include scenarios with high and very high GHG emissions (SSP3-7.0 and SSP5-8.5) and CO_2 emissions that roughly double from current levels by 2100 and 2050, respectively; scenarios with intermediate GHG emissions (SSP2-4.5) and CO_2 emissions remaining around current levels until the middle of the century; and scenarios with very low and low GHG emissions and CO_2 emissions declining to net zero around or after 2050, followed by varying levels of net negative CO_2 emissions (SSP1-1.9 and SSP1-2.6).

IPCC 报告《气候变化 2021》：物理科学基础评估了 5 种可能的未来发展情景和相应的气候影响。情景评估从 2015 年开始，其中包括温室气体排放量高和非常高的情景（SSP3-7.0 和 SSP5-8.5）以及到 2100 年和 2050 年二氧化碳排放量分别比当前水平几乎翻倍的情景；中等级别温室气体排放情景（SSP2-4.5）直到 21 世纪中期，二氧化碳排放量始终保持在当前水平；温室气体排放量极低的情景和二氧化碳排放量在 2050 年前后降至净零，随后出现不同水平的负二氧化碳排放量情景（SSP1-1.9 和 SSP1-2.6)。

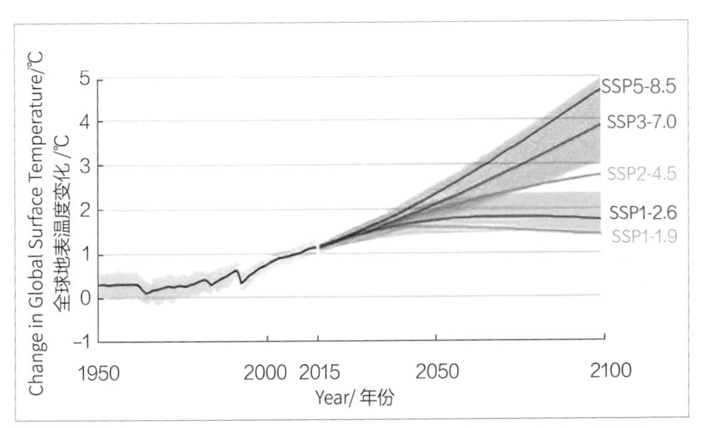

Global Surface Temperature Change Relative to 1850—1900
全球地表温度变化（相较于 1850—1900 年）

Source 资料来源：IPCC 联合国政府间气候变化专门委员会。

Global surface temperature will continue to increase until at least mid-century under all emissions scenarios considered. Global warming of 1.5°C and 2°C will be exceeded during the 21st century unless deep reductions in CO_2 and other greenhouse gas emissions occur in the coming decades.

在所有纳入考量的排放情景下，全球地表温度至少将持续上升到 21 世纪中期。除非未来几十年二氧化碳和其他温室气体排放量大幅减少，否则 21 世纪的全球升温幅度将超过 2℃。

*SSP：Shared Socioeconomic Pathways 共享社会经济路径。

Predictions of Future Climate Change
气候变化的预测

A global surface temperature change of 1.5°C is among the most optimistic scenarios. Yet, it will still significantly affect our climate indicators.

全球地表温度上升 1.5℃是最乐观的情景之一。然而，它仍将对关键气候指标产生显著影响。

Sea level 海平面	0.26 to 0.77 m global mean sea level rise (relative to 1986—2005) 全球平均海平面上升 0.26 ~ 0.77 米（相较于 1986—2005 年）
Ocean heat and acidification 海洋升温和酸化	Coral reefs are projected to decline by a further 70%~90% Global annual catch for marine fisheries will decrease by 1.5 million tonnes 珊瑚礁预计将进一步减少 70%~90% 全球海洋渔业年捕捞量将减少 150 万吨
Temperature anomalies 温度异常	3℃ hotter extreme hot days in mid-latitudes 4.5℃ warmer extreme cold nights in high latitudes 中纬度地区极端高温天的气温将上升 3℃ 高纬度地区极端寒冷夜晚的气温将上升 4.5℃

Carbon emissions leading to a 2°C temperature change is the greatest tolerable change. From the information given above, try to guess how this scenario will affect our climate indicators. (See next page for answers)

导致气温上升 2℃的碳排放是地球可承受的最大限度。根据上面的信息，尝试猜测在此情景下的关键气候指标将受到什么样的影响。（答案见下页）

Sea level 海平面	______ to ______ m global mean sea level rise (relative to 1986—2005) 全球平均海平面上升 ______ 至 ______ 米（相较于 1986—2005 年）
Ocean heat and acidification 海洋升温和酸化	Coral reefs are projected to decline by ______ % Global annual catch for marine fisheries will decrease by ______ million tonnes 珊瑚礁预计将减少 ______ % 全球海洋渔业年捕捞量将减少 ______ 万吨
Temperature anomalies 温度异常	______ ℃ hotter extreme hot days in mid-latitudes ______ ℃ warmer extreme cold nights in high latitudes 中纬度地区极端高温天的气温将上升 ______ ℃ 高纬度地区极端寒冷夜晚的气温将上升 ______ ℃
Others 其他	Higher risks from heavy precipitation, droughts, and forest fires Risks from some vector-borne diseases, such as malaria and dengue fever 大量降雨、干旱和森林火灾带来更高风险 疟疾和登革热等媒介传播疾病的风险

Source 资料来源： IPCC 联合国政府间气候变化专门委员会。

Predictions of Future Climate Change
气候变化的预测

Many changes in the climate system become larger in direct relation to increasing global warming. They include increases in the frequency and intensity of hot extremes, marine heatwaves, heavy precipitation, and, in some regions, agricultural and ecological droughts; an increase in the proportion of intense tropical cyclones; and reductions in Arctic sea ice, snow cover and permafrost.

气候系统的许多变化与全球变暖的加剧有直接关系，包括极端高温、海洋热浪、强降水的频率和强度增加，以及在某些地区的农业和生态干旱；强烈热带气旋的比例增加；北极海冰、积雪和永久冻土的减少。

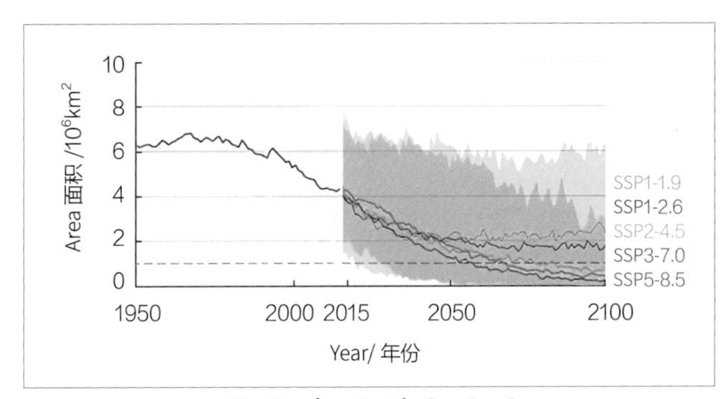

September Arctic Sea Ice Area
九月的北极海冰面积

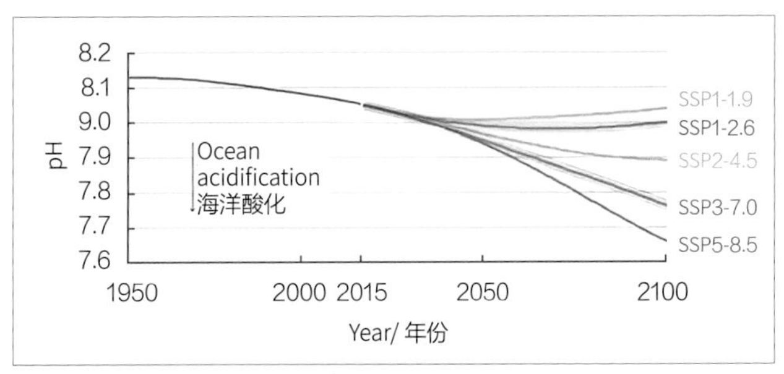

Global Ocean Surface pH
全球海洋表面 pH

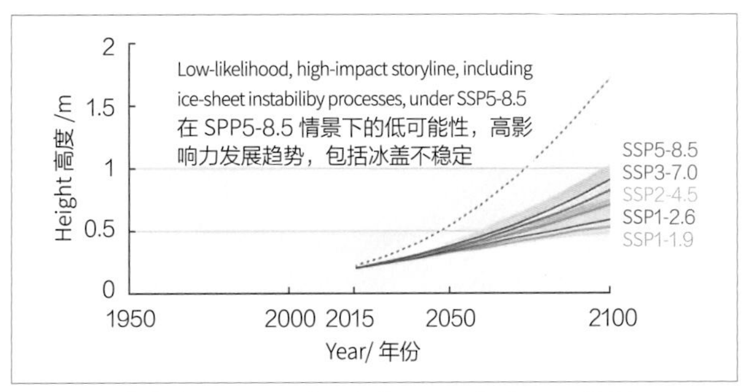

Global mean sea level change relative to 1900
全球平均海平面变化（相较于 1900 年）

Source 资料来源：IPCC 联合国政府间气候变化专门委员会。

Additional warming is projected to further amplify permafrost thawing and loss of seasonal snow cover, of land ice and of Arctic sea ice. The Arctic is likely to be practically sea ice-free in September 31 at least once before 2050 under the five illustrative scenarios considered in this report, with more frequent occurrences for higher warming levels.

Over the rest of the 21st century, likely ocean warming ranges from 2～4 (SSP1-2.6) to 4～8 times (SSP5-8.5) the 1971—2018 change.

It is virtually certain that global mean sea level will continue to rise over the 21st century.

Global mean sea level rise above the likely range-approaching 2m by 2100 and 5 m by 2150 under a very high GHG emissions scenario (SSP5-8.5) cannot be ruled out due to deep uncertainty in ice-sheet processes.

气温持续变暖将进一步使永久冻土融化范围以及季节性积雪、陆地冰和北极海冰的损失扩大。根据 IPCC 的报告中的 5 种情景，在 2050 年之前，北极在 9 月 31 日可能至少出现一次几乎没有海冰的情况。升温水平越高，这种情况就越频繁。

在 21 世纪剩下的时间里，海洋的变暖可能会从 1971—2018 年的 2～4 倍 (SSP1-2.6) 增长到 4～8 倍 (SSP5-8.5)。

几乎可以肯定的是，全球平均海平面将在 21 世纪继续上升。

因为冰盖发展趋势存在高度不确定性，所以不能排除全球平均海平面上升值超过预期范围的可能性。在温室气体排放量非常高的情景下 (SSP5-8.5)，到 2100 年全球平均海平面可能上升接近 2 米，到 2150 年上升接近 5 米。

Answer Key
答案（P.15）：
1. 0.30，0.83
2. 99，3；99，300
3. 4，6

MULTIPLE DIMENSIONS OF CLIMATE CHANGE IMPACT

气候变化的多元影响

Climate Change Potential Consequences
气候变化的潜在影响

If greenhouse gas concentrations keep rising, climatic changes are likely to result. Those changes will potentially have wide-ranging effects on the environment and socio-economic and related sectors, such as health, agriculture, forests, water resources, coastal areas and biodiversity.

如果温室气体浓度持续上升，气候变化很可能会加剧。这些变化可能会对环境、社会经济及相关部门，例如健康、农业、森林、水资源、沿海地区和生物多样性，产生广泛影响。

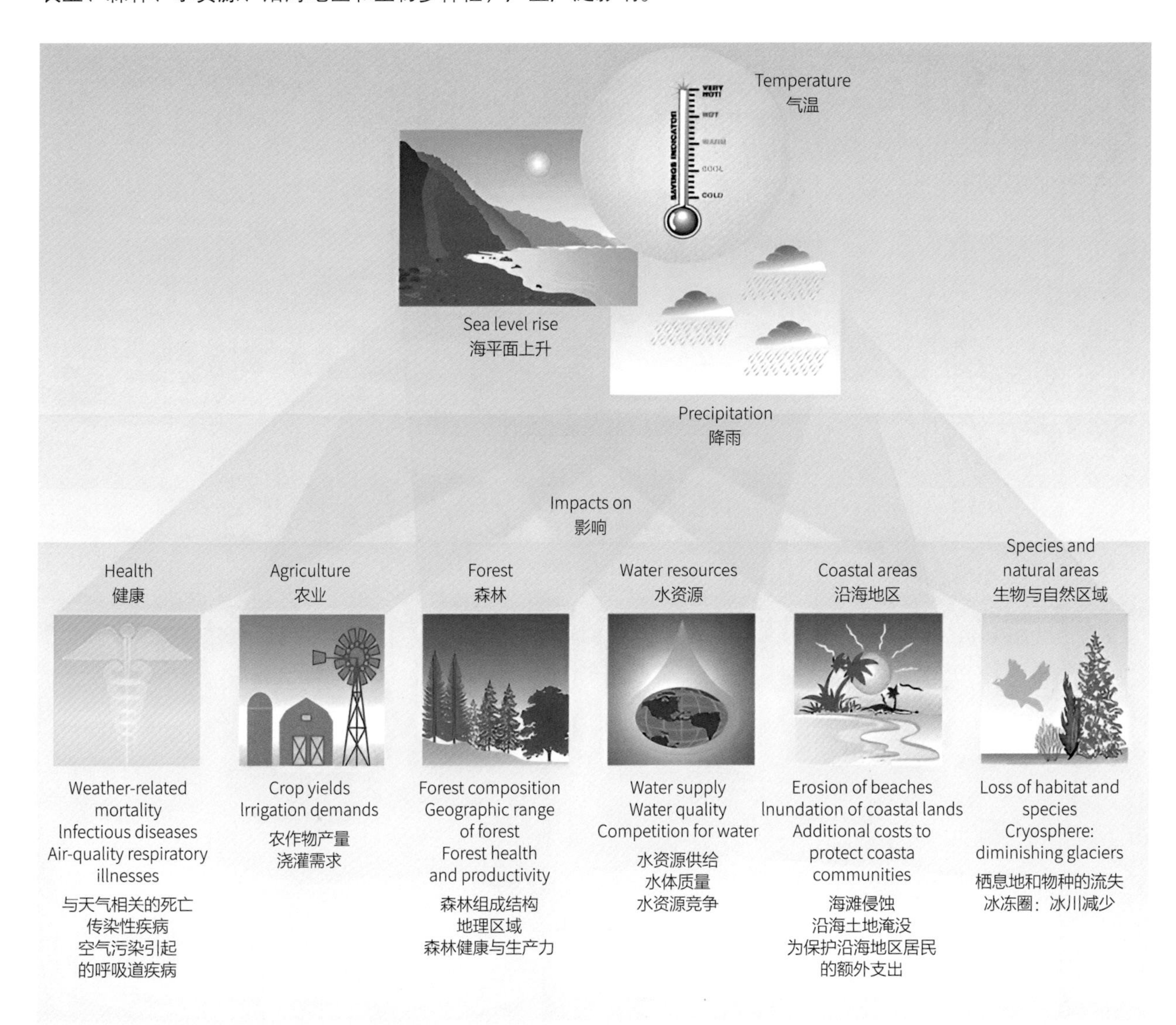

Source 资料来源：GRID-Arendal 全球资源信息数据库–阿伦达尔中心。

Choose one of the themes from above, do some research and make a poster about the local impact of climate change on the theme you have chosen.

从上图中的几个主题中选择一个，做一些调查后，制作一张海报，以说明气候变化对你所在地区的影响。

Environmental Impact of Climate Change
气候变化的环境影响

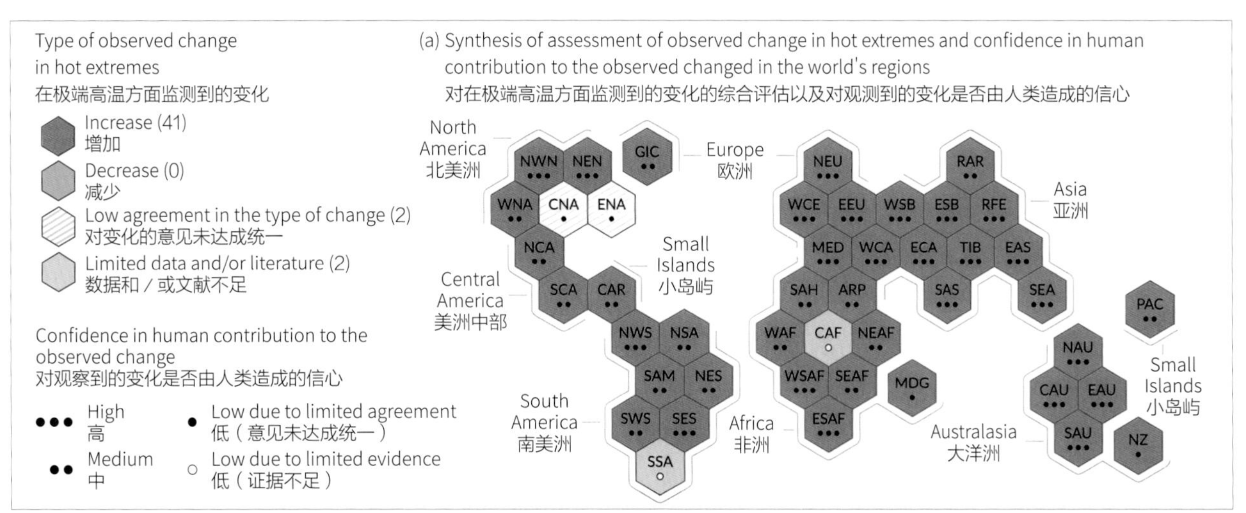

Type of observed change since the 1950s/ 自 1950 年以来观测到的变化

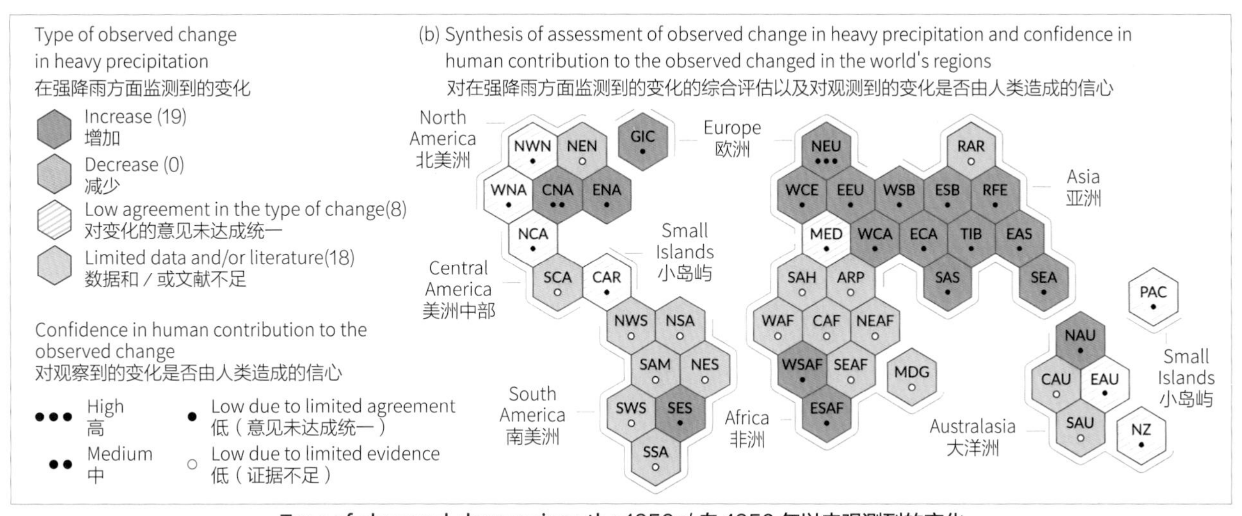

Type of observed change since the 1950s/ 自 1950 年以来观测到的变化

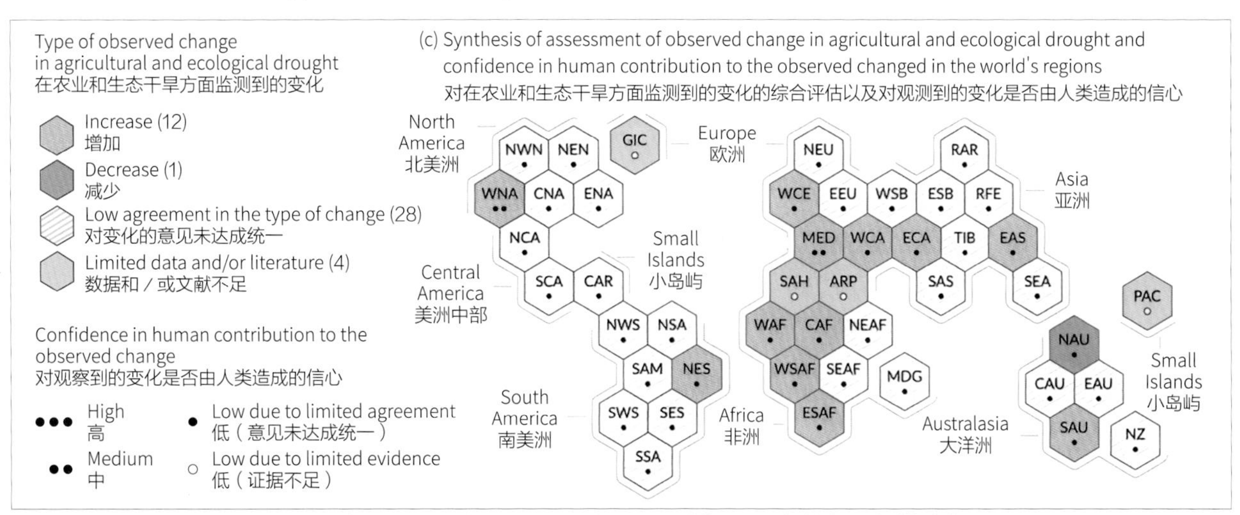

Type of observed change since the 1950s/ 自 1950 年以来观测到的变化

Source 资料来源：IPCC 联合国政府间气候变化专门委员会。

Environmental Impact of Climate Change
气候变化的环境影响

Loss in Biodiversity/ 生物多样性丧失

According to WWF's Living Planet Report 2020, climate change adversely affects genetic variability, species richness and populations, and ecosystems. In turn, loss of biodiversity can adversely affect climate-for example, deforestation increases the atmospheric abundance of carbon dioxide.

As temperatures change, some species will need to adapt by shifting their range to track a suitable climate. The effects of climate change on species are often indirect. Changes in temperature can confound the signals that trigger seasonal events such as migration and reproduction, causing these events to happen at the wrong time (for example misaligning reproduction and the period of greater food availability in a specific habitat).

根据 WWF 的《地球生命力报告 2020》，气候变化会对基因多样性、物种丰富度和生态系统产生诸多不利影响。生物多样性丧失会对气候产生负面影响，例如，砍伐森林会增加大气中的二氧化碳含量。

随着温度的变化，一些物种将需要通过改变它们的活动范围来适应气候。气候变化对物种的影响通常是间接的。温度变化会混淆触发迁移和繁殖等季节性事件的信号，使这些事件发生在错误的时间（例如，繁殖期错位、特定栖息地中食物供应量增加的时期改变）。

Changes in Ocean Circulation Patterns and Productivity
海洋环流模式和生产力的变化

Species Moving Away from Warming Waters
物种搬离变暖水域

Changes in Ecological Interactions and Metabolism
生态相互作用和新陈代谢的变化

Changes in Disease Incidence
发病率的变化

Coral Bleaching
珊瑚白化

Change in Timing of Biological Processes
生物过程的时间变化

Polar bears are one of the most affected species of climate change, their dependence on sea ice makes them highly vulnerable to a changing climate. Polar bears rely heavily on the sea ice environment for traveling, hunting, mating, resting, and in some areas, maternal dens. In particular, they depend heavily on sea ice-dependent prey, such as ringed and bearded seals. Additionally, their long generation time and low reproductive rate may limit their ability to adapt to changes in the environment.

北极熊是受气候变化影响最严重的物种之一，对海冰的依赖使它们极易受到气候变化的影响。北极熊在很大程度上依赖海冰进行活动、狩猎、交配、休息，并且在某些海冰上建立母巢。特别是它们主要捕食那些依赖于海冰生存的猎物，例如，环斑海豹和胡须海豹。此外，由于它们的生存时间较长、繁殖率较低，它们适应环境变化的能力会受到极大限制。

Source 资料来源：WWF 世界自然基金会。

Environmental Impact of Climate Change
气候变化的环境影响

Coral Bleaching/ 珊瑚白化

Coral reefs occur in more than 100 countries and territories and whilst they cover only 0.2% of the seafloor, they support at least 25% of marine species and underpin the safety, coastal protection, food and economic security of hundreds of millions of people.

However, coral reefs are among the most vulnerable ecosystems on the planet to anthropogenic pressures, including global threats from climate change and ocean acidification. According to the National Oceanic and Atmospheric Association, more than 75% of Earth's tropical reefs experienced bleaching-level heat stress between 2014 and 2017, and for 30% of reefs, it reached mortality level.

珊瑚礁出现在 100 多个国家和地区，虽然它们仅覆盖了 0.2% 的海底，但对至少 25% 的海洋物种的生存是至关重要的。它们同时还能保障数亿人的安全，保护海岸，提供食物，促进经济发展。

然而，珊瑚礁是地球上最容易受到人为因素影响的生态系统之一，其影响因素包括气候变化和海洋酸化带来的威胁。根据美国国家海洋和大气协会的数据，2014—2017 年，地球上超过 75% 的热带珊瑚礁经历了会导致其白化的等级的热浪， 这对 30% 的珊瑚礁是致命的。

A warming ocean
causes thermal stress that contributes to coral bleaching and infectious disease
海洋变暖
引起热应力，导致珊瑚白化并感染传染病

Ocean acidification as a result of increased CO_2
a reduction in pH levels decreases coral growth and structural integrity
二氧化碳增加所导致的海洋酸化
海水 pH 降低延缓珊瑚生长和结构的完整性

Changes in storm patterns
stronger and more frequent storms can cause the destruction of coral reefs
风暴规律的变化
更强烈、频繁的风暴可能会破坏珊瑚礁

Altered ocean currents
leads to changes in temperature regimes that contribute to lack of food for corals
洋流变化
温度变化导致珊瑚缺乏食物

Sea level rise
may lead to increases in sedimentation for reefs located near land-based sources of sediment, increasing the risk of smothering of coral
海平面上升
可能会导致位于陆地附近的珊瑚礁面临沉积物增加的风险，沉积物可能导致珊瑚窒息

Changes in precipitation
increased runoff of freshwater, sediment, and land-based pollutants contribute to algal blooms and cause murky water conditions that reduce light
降水变化
淡水径流、沉积物和陆基污染物的增加会导致藻类大量繁殖并使水体浑浊，进而光照减少

Climate change is a primary threat to the health of the Australia Great Barrier Reef. Ocean warming has triggered the third global bleaching event, which could wipe out all the coral reefs in the north-central part of the reef, accounting for more than 40% of the entire Great Barrier Reef.

气候变化是对澳大利亚大堡礁的生态系统健康危害最大的因素。海洋变暖引发的第三次全球白化事件，导致中北部的珊瑚礁可能被全数摧毁，其数量占据大堡礁整体的 40%以上。

Source 资料来源：Global Coral Reef Monitoring Network 全球珊瑚礁监测网络；National Oceanic& Atmospheric Administration (NOAA) 美国国家海洋大气局。

Economic Impact of Climate Change
气候变化的经济影响

Financial Risk of Climate Change / 气候变化的金融风险

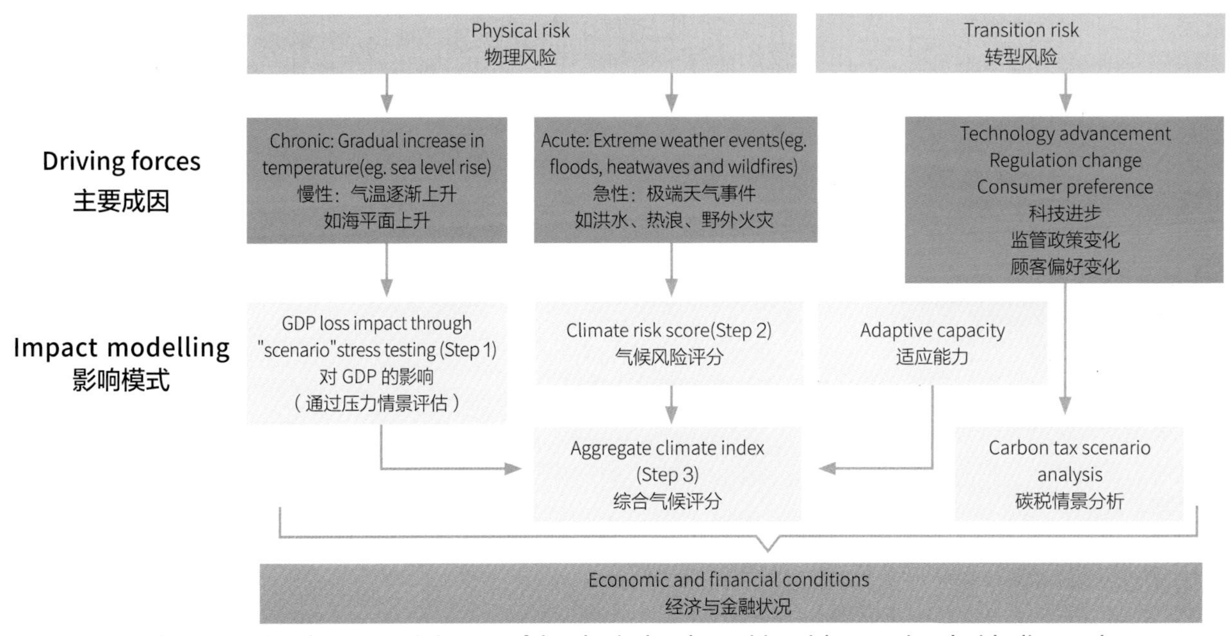

Approach to assessing the economic impact of the physical and transition risks associated with climate change
评估与气候变化相关的物理风险和转型风险的经济影响方法

The expected impact on global GDP by 2050 under three different scenarios compared to a world without climate change:
与没有发生气候变化的情景相比，在以下 3 种情境下，预计到 2050 年，全球 GDP 将受到以下影响：

–4%

If Paris Agreement targets are met
(a well-below 2°C increase)
如果达到《巴黎协定》的目标
（远低于 2℃的气温升高）

–11%

If further mitigating actions are taken
(2°C increase)
如果采取进一步的缓解措施
（气温升高 2℃）

–18%

If no mitigating actions are taken
(3.2°C increase)
如果不采取缓解措施
（气温升高 3.2℃）

GDP measures the monetary value of final goods and services - that is, those that are bought by the final user-produced in a country in a given period of time (a quarter or a year). (IMF, 2020)

GDP 衡量了一个国家在给定时间段（一个季度或一年）内生产的最终商品和服务，即最终消费者购买的那些商品和服务的货币价值。（国际货币基金组织，2020）

Source 资料来源：Swiss Re Institute 瑞士再保险研究院。

Economic Impact of Climate Change
气候变化的经济影响

Economic Risk of Climate Change / 气候变化的经济风险

The economic impact of climate change will be different for different regions and industries
气候变化对不同地区和行业的经济影响会有所不同

Transition risk 转型风险

The transition to a low carbon economy carries both risk and opportunity and could unfold gradually over time or through sudden shocks. Transition risks include policy changes, reputational impacts, and shifts in market preferences, norms and technology.

The diagram below illustrates the effect on the revenue of different industries in different regions, as a result of the imposition of a global carbon tax of USD 100 per metric ton. The energy, materials and utilities sectors will be the most impacted.

向低碳经济的转型既有风险也有机遇，这些风险可能是缓慢发生的，也可能是突然的冲击。转型风险包括政策变化，声誉影响，以及市场偏好、行为规范和技术的变化。

下图展现了若征收 100 美元 /tCO_2 当量全球碳税，将对不同地区的各行业的收入造成的影响，其中能源、原材料和公用设施行业将受到最大冲击。

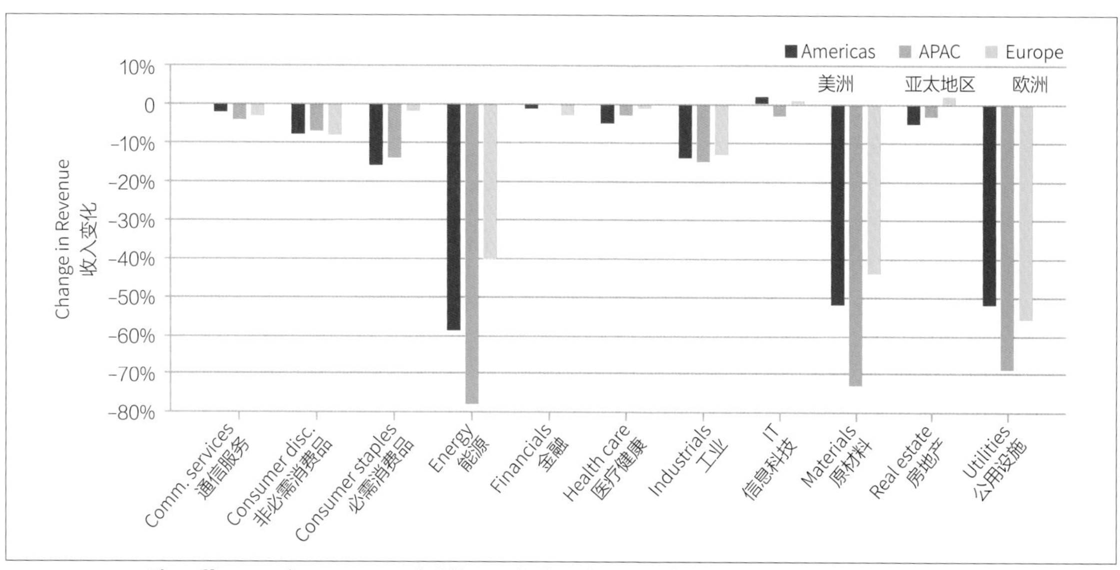

The effect on the revenue of different industries in different regions, as a result of the imposition of a global carbon tax of USD 100 per metric ton

若征收每 1 吨二氧化碳当量 100 美元的全球碳税，将对不同地区的各行业的收入造成的影响

Discussion
讨论议题

What is the difference between transition risks and physical risks, can you give some examples of each?

Which kind of risk do you think is more detrimental to our economy? Why?

Why do you think the energy, materials and ultilities sector are the most impacted? What are some features that makes a sector more likely to be affected by climate change risks?

转型风险和物理风险之间有什么区别，你能举一些例子吗?

你认为哪种风险对我们的经济影响更大? 为什么?

你认为为何能源、原材料等行业受到的影响最大? 哪些特征使一个行业更容易受到气候变化风险的影响?

Source 资料来源：Swiss Re Institute，瑞士再保险研究院；Cambridge Institute for Sustainability Leadership (CISL) 剑桥可持续发展领导力学院。

Social Impact of Climate Change
气候变化的社会影响

Climate Change Health Consequences/ 气候变化的健康影响

Direct Impact 直接影响	Consequences on Human Health 对人类健康的影响
Severe Weather 恶劣天气	Injuries, fatalities, mental health impacts 受伤、意外死亡、心理健康影响
Air Pollution 空气污染	Asthma, Cardiovascular Disease 哮喘、心血管疾病
Changes in Vector Ecology 生态媒介变化	Malaria, dengue, encephalitis, hantavirus, Rift Valley fever, Lyme disease, chikungunya, West Nile virus 疟疾、登革热、脑炎、裂谷热、莱姆病、基孔肯雅热、西尼罗河病毒
Increasing Allergens 过敏原增加	Respiratory allergies, asthma 呼吸道过敏、哮喘
Water Quality Impacts 水质量影响	Cholera, cryptosporidiosis, campylobacter, leptospirosis, harmful algal blooms 霍乱、隐孢子虫病、弯曲杆菌、钩端螺旋体病、有害藻华
Water and Food Supply Impacts 水与食物供给影响	Malnutrition, diarrhoeal disease 营养不良、腹泻
Environmental Degradation 环境退化	Forced migration, civil conflict, mental health impacts 强制移民、社会冲突、心理健康影响
Extreme Heat 极端炎热	Heat-related illness and death, cardiovascular failure 炎热相关疾病与死亡、心血管疾病

Climate change is the single biggest health threat facing humanity. The impacts are already harming health through air pollution, disease, extreme weather events, forced displacement, food insecurity and pressures on mental health.

The World Health Organization (WHO) reports that climate change is responsible for at least 150,000 deaths per year, a number that is expected to double by 2030. The effects of global warming will cause dire health consequences, including: Infectious diseases, Heat Strokes, Asthma and other respiratory diseases, loss of agriculture productivity.

气候变化是人类面临的最大健康威胁，它通过空气污染、疾病传播、极端天气、被迫移居、粮食供应不稳定和心理健康压力而危害人类健康。

世界卫生组织 (WHO) 报告称，气候变化每年导致至少 150 000 人死亡。预计这个数字到 2030 年将翻 1 倍。全球变暖将导致严重的健康问题，包括传染病、中暑、哮喘和其他呼吸道疾病。气候变化还会降低农业生产力。

Try to recall whether you or someone close to you has been sick in the past one year, do you think it has something to do with climate change? Why?

试回想你或身边的人在过去一年中有没有生过病？你认为它与气候变化有关吗？为什么？

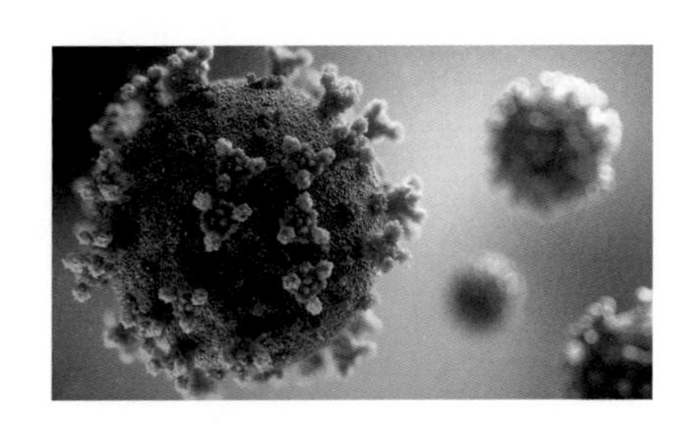

Social Impact of Climate Change
气候变化的社会影响

Climate Justice/ 气候正义

"Climate change is happening now and to all of us. No country or community is immune, and, as is always the case, the poor and vulnerable are the first to suffer and the worst hit."

——UN Secretary-General António Guterres

"气候变化正发生在我们所有人身上。没有一个国家或社区可以幸免，而穷人和弱势群体是最先受害并且受害最严重的人群。"

——联合国秘书长安东尼奥 · 古特雷斯

Climate justice "insists on a shift from a discourse on greenhouse gases and melting ice caps into a civil rights movement with the people and communities most vulnerable to climate impacts at its heart."

——Mary Robinson, Former President of Ireland

气候正义"主张从关于温室气体和冰盖融化的议题延伸到以最容易受到气候影响的人民和社区为中心的民权运动。"

——爱尔兰前总统玛丽 · 罗宾逊

As climate change disproportionately affects ecosystems and natural resources, infrastructure and human settlements, livelihoods, and health and security, it directly and indirectly affects the enjoyment of many of our human rights. It is one of the greatest threats to the enjoyment of human rights for present and future generations, with a disproportionate impact on vulnerable individuals, groups and peoples. Some groups are more likely to be harmed by climate change, including women, children, future generations, indigenous peoples, environmental human rights defenders, migrants, persons with disabilities, older persons and the poor. States should pay special attention to these vulnerable groups when planning and implementing measures to address climate change.

气候变化严重影响了生态系统和自然资源、基础设施、人类的居住环境、生存方式、健康、安全等，直接和间接对人权造成了进一步的影响，是对当代和后代人享有人权的最大威胁之一。对弱势个人或群体，包括妇女、儿童、后代、土著居民、环境保护主义者、移民、残疾人、老年人和穷人，气候变化的影响更为严重。因此，各国在规划和采取应对气候变化的措施时，应该对上述人群给予特别关注。

Human Rights
人权

Human rights are rights inherent to all human beings without discrimination, regardless of nationality, place of residence, sex, national or ethnic origin, colour, religion, language, sexual orientation and gender identity, disability, or any other status. These include: civil and political rights, such as the right to life, equality before the law and freedom of expression; economic, social and cultural rights, such as the rights to work, social security and education; or collective rights, such as the rights to development and self-determination.

The *Universal Declaration of Human Rights*, adopted by the United Nations General Assembly in 1948, was the first legal instrument to enshrine fundamental human rights that should be universally protected. It contains 30 articles that provide principles for current and future human rights conventions, treaties, and other legal instruments and basic elements.

人权是我们与生俱来的权利，不分国籍、居住地、性别、民族或人种、肤色、宗教、语言、性取向和性别认同、残疾或其他任何情况。它们涵盖的范围非常广泛，从最基本的生命权，到食物、教育、工作、健康和人身自由等权利。

联合国大会于 1948 年通过的《世界人权宣言》是第一份载明人权应受到普遍保护的基本权利的法律文书。其中所记载的 30 项条款为当前和今后的人权公约、条约以及其他法律文书提供了基本原则。

Social Impact of Climate Change
气候变化的社会影响

Climate Justice/ 气候正义

Human Right to Water and Sanitation 享有水和卫生设施的人权

Climate change impedes access to safe and clean drinking water and sanitation for all. Climate change is projected to reduce renewable surface water and groundwater resources in most dry subtropical regions intensifying competition for water. It will likely increase the risk of water scarcity in urban areas whilst rural areas are expected to experience major impacts on water availability and supply.

气候变化将减少大部分干燥的亚热带地区的可再生地表水和地下水资源，从而加剧这些地区对水资源的竞争。水资源减少可能会增加城市地区的缺水风险，而农村地区的水供应也会受到严重影响，这阻碍了所有人获得安全、清洁的饮用水和卫生设施的权利。

Human Right to Health 享有健康的人权

Climate change hampers the enjoyment of the right to health as its impacts cause a greater risk of injury and disease, for example due to more intense fires, heatwaves and heat-amplified levels of smog that could exacerbate respiratory disorders. There is also an increased risk of malnutrition and undernutrition resulting from diminished food production in some regions and an increased risk of vector-borne diseases.

气候变化会导致受伤和患病的风险增加，例如更强烈的火灾、热浪和因高温加剧的空气污染物所导致的呼吸系统疾病，威胁着人们的健康权。同时，由于某些地区粮食减产和传播疾病的媒介增加，营养不良的风险也相应加剧。

Human Right to Housing 享有住房的人权

Climate change effects such as extreme weather events infringe on the right to housing by destroying homes and displacing multitudes of people. Over time, drought, erosion and flooding leave territories inhabitable resulting in displacement and migration, and sea level rise also threatens the land upon which houses in low-lying areas are situated. Climate and weather-related disasters alone contribute to the displacement of more than 20 million people each year.

部分因气候变化导致的极端天气事件会破坏房屋，导致大量人口流离失所，威胁了人们的住房权。随着时间的流逝，干旱和洪水冲蚀使许多地区逐渐无法居住，人们流离失所或被迫移民。同时，海平面上升对低洼地区的房屋也造成威胁。仅气候和天气相关的灾害每年就导致超过2 000万人流离失所。

Human Right to Food 享有食物的人权

Climate change undermines the right to food as climate induced droughts and floods, as well as unpredictable rain patterns, lead to food insecurity particularly in regions reliant on rainfed agriculture. Further, the acidification of oceans due to greater carbon dioxide concentration in the atmosphere destroys coral reefs leading to a decrease in fish stocks.It is estimated that climate change could render 600 million more people vulnerable to malnutrition by 2080.

气候变化引起的干旱和洪水以及不可预测的降雨模式会导致粮食供应不稳定，特别是依赖于雨养农业的地区。此外，大气中二氧化碳浓度增加使海洋酸化，破坏珊瑚礁，进而导致鱼类种群减少。据估计，到 2080 年，气候变化可能会导致 6 亿多人面临营养不良的风险。

Social Impact of Climate Change
气候变化的社会影响

Climate Justice/ 气候正义

Human Right to Culture
享有文化的人权

Indigenous peoples have the right to maintain, control, protect and develop their cultural heritage, traditional knowledge and traditional cultural expressions. Climate change impacts pose a threat to the enjoyment of cultural rights by interfering with cultural practices and meaningful spaces for cultural interactions and hampering the continuity of specific practices and ways of life.

土著居民有权维护和发扬其文化遗产、传统知识和传统文化表现形式。气候变化影响文化习俗和有意义的文化互动空间，阻碍特定习俗和生活方式的传承，从而对文化权构成威胁。

Human Right to Education
享有教育的人权

Extreme weather events lead to food insecurity, reduce the availability of safe drinking water, compromise sanitation and increase the incidence of weather-related diseases such as malaria and diarrheal diseases, leading to absenteeism and possible withdrawal of children from school, thus infringing on these children's right to education.

极端天气会导致粮食供应失去保障，减少安全饮用水的供应，影响卫生条件并增加疟疾和腹泻等与天气有关的疾病的发病率，还导致儿童可能无法正常上课，甚至是辍学，从而侵犯了这些儿童受教育的权利。

Activity: Human Right to a Healthy Environment
活动： 享有健康环境的人权

In 2021, the Human Rights Council declared for the first time that having a clean, healthy and sustainable environment is a human right. The United Nations Environment Programme points out that air pollution and climate change are closely interlinked. Air pollutants include more than just greenhouse gases—principally carbon dioxide but also methane, nitrous oxide and others—but there's a big overlap. For instance, air pollution in the form of particulate matter from diesel engines is circulated around the globe, ending up in the most remote places, including the polar regions. When it lands on ice and snow it darkens them slightly, leading to less sunlight being reflected back into space, and contributing to global warming.

2021 年，人权理事会首次宣布，拥有一个清洁、健康和可持续的环境是一项人权。 联合国环境规划署指出，空气污染和气候变化紧密关联。空气污染物不仅仅包括二氧化碳，也包含甲烷、一氧化二氮等温室气体，两者有很大一部分是重叠的。例如，柴油发动机产生的颗粒物所造成的空气污染会在全球传播，最终到达最偏远的地方，包括极地地区。当它们触及冰与雪时，会使冰与雪的颜色变暗，导致反射回太空的阳光减少，加剧全球变暖。

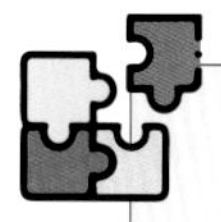

Discuss with your classmates, to what extent do you think the right to a healthy environment has been realized in your country or region? Create a poster to raise awareness and advocate for the right to a healthy environment.

与同学讨论，你认为你所在的国家或地区在多大程度上实现了健康环境权?
制作一张海报以提高公众意识并提倡健康环境权。

Source 资料来源：UNEP 联合国环境规划署。

CLIMATE CHANGE AGREEMENTS AND POLICY FRAMEWORKS

气候变化协议与政策框架

Stockholm Conference
斯德哥尔摩会议

The Stockholm Conference was the first world conference to make the environment a major issue. The participants adopted a series of principles for sound management of the environment including the *Stockholm Declaration* and *Action Plan for the Human Environment* and several resolutions. The *Stockholm Declaration*, which contained 26 principles, placed environmental issues at the forefront of international concerns and marked the start of a dialogue between industrialized and developing countries on the link between economic growth, the pollution of the air, water, and oceans and the well-being of people around the world.

1972年，联合国人类环境会议在斯德哥尔摩举行，这是首届将环境问题作为主要议题的世界会议。会议通过了一系列环境管理原则，包括《斯德哥尔摩宣言》《人类环境行动计划》以及若干决议。《斯德哥尔摩宣言》中记载的26项原则将环境问题置于国际关注的首要位置，这象征着工业化国家与发展中国家开始就经济增长、空气、水和海洋的污染以及全世界人民福祉之间的关联展开对话。

Stockholm Declaration and Action Plan for the Human Environment
《斯德哥尔摩宣言》与《人类环境行动计划》

Global Environmental Assessment Programme (watch plan) Activities 全球环境评估方案（观察计划）活动	Environmental Management 环境管理活动	International Measures to Support Assessment and Management Activities Carried Out at the Nnational and International Levels 支援在国家层面和国际层面开展评估和管理活动的国际措施

The United Nations Environment Programme 联合国环境规划署 (UNEP)

A significant achievement of the Conference was the creation of UNEP.
会议的主要成果之一是建立了联合国环境规划署（以下简称环境署）。

UNEP's mission is to provide leadership and encourage partnership in caring for the environment by inspiring, informing, and enabling nations and peoples to improve their quality of life without compromising that of future generations.

UNEP employs seven interlinked subprogrammes for action: Climate Action, Chemicals and Pollutions Action, Nature Action, Science Policy, Environmental Governance, Finance and Economic Transformations and Digital Transformations.

环境署的使命是激励、推动和促进各国及其人民在不损害子孙后代生活质量的前提下提高自身的生活质量，为建立关爱环境的伙伴关系提供指导和鼓励。

环境署采用七个相互关联的行动子方案：气候行动、化学品和污染行动、自然行动、科学政策、环境治理、金融和经济转型以及数字转型。

Source 资料来源：UNEP 联合国环境规划署。

UNFCCC-United Nations Framework Convention on Climate Change 《联合国气候变化框架公约》

UNFCCC is a result of global consensus that we need to combat climate change together
UNFCCC 是全球就应对气候变化达成共识的结果

Climate change is a common concern of humankind 气候变化是人类共同面临的危机	GHG emissions contribute to climate change irrespective of their origin 无论哪国排放的温室气体都会影响气候变化	All countries will be affected if no action is taken 如果不采取行动，所有国家都将受到影响	A global agreement is needed to regulate emissions and help countries to adapt 需要达成全球共识来规范排放并帮助各国适应气候变化

1979年 First World Climate Conference 第一届世界气候大会

This was the first time that scientific evidence on the influence of human activities on the global climate was presented at an international level. Governments concluded that there was a need for an impartial and independent body to address this issue, leading to the creation of the Intergovernmental Panel on Climate Change (IPCC) in 1988.

1979 年，第一届世界气候大会在日内瓦召开，这是人类活动对全球气候影响的科学证据首次在国际层面上亮相。各国政府的结论是需要一个公正、独立的机构来解决这个问题，因而促成了成立于 1988 年的联合国政府间气候变化专门委员会 (IPCC)。

> The framework convention sets out basic obligations of all "Parties" to combat climate change
>
> 《框架公约》规定了所有缔约方在应对气候变化问题上的基本义务

1990年 IPCC First Assessment Report IPCC 第一次评估报告

In 1990, the IPCC First Assessment Report was published and the Second World Climate Conference took place, leading to a ministerial declaration calling for a global treaty to address the problem of human induced climate change.

1990 年，IPCC 发布了第一次评估报告，并召开了第二次世界气候大会，促成了一项部长级宣言，呼吁制定一项全球条约来解决人为引起的气候变化问题。

> Currently has 197 Parties, including 196 states and 1 regional organization(the EU)
>
> 目前有 197 个缔约方，其中包括 196 个国家和 1 个区域组织（欧盟）

1992年 UN Framework Convention 联合国框架公约

The *United Nations Framework Convention on Climate Change* (UNFCCC) was adopted as a basis for a global response to the climate change problem.

《联合国气候变化框架公约》(UNFCCC)(以下简称《框架公约》) 被采纳成为全球应对气候变化问题的基础。

> In the context of climate negotiations, Parties work through different Party groupings that best represent their interests
>
> 在气候谈判的背景下，各缔约方通过最能代表其利益的不同缔约方群组开展工作

Source 资料来源：UNFCCC《联合国气候变化框架公约》。

UNFCCC-United Nations Framework Convention on Climate Change
《联合国气候变化框架公约》

The Ultimate Aim of the Convention
《框架公约》的最终目标

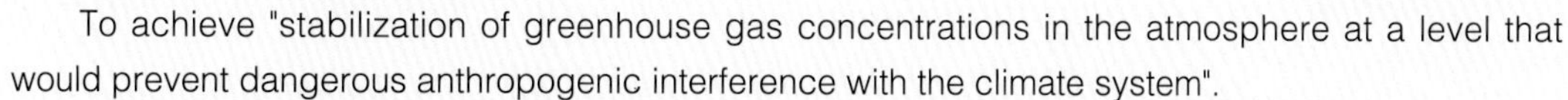

To achieve "stabilization of greenhouse gas concentrations in the atmosphere at a level that would prevent dangerous anthropogenic interference with the climate system".

为了实现“将大气中温室气体的浓度稳定在防止气候系统受到危险的人为干扰的水平上”的目标。

Such a level should be achieved within a time frame sufficient to:

- allow ecosystems to adapt naturally to climate change;
- ensure that food production is not threatened;
- enable economic development to proceed in a sustainable manner.

在一定时间范围内：

- 使生态系统能够自然地适应气候变化；
- 确保粮食生产不受威胁；
- 使经济能够可持续地发展。

Mitigation and adaptation are the two broad responses adopted by the Convention to address climate change
减缓和适应是《框架公约》通过的为应对气候变化而采取的两大措施

Mitigation 减缓

Mitigation of climate change and its impacts, through stabilization of GHG emissions, is the essence of the Convention's objective

通过稳定温室气体净排放量来减轻气候变化及其影响是《框架公约》目标的精髓

Adaptation 适应

Actions to help adapt to climate conditions and climate impacts

帮助适应气候条件和气候影响的行动

Match the following Guiding Principles of the Convention with its definition. (See next page for answers)
试将以下《框架公约》的指导性原则与其定义匹配。（答案见下页）

A. The lack of scientific certaitny should not prevent Parties from taking measures 缺乏科学确定性不应妨碍缔约方采取措施	1. Right to Sustainable Development 可持续发展的权利
B. Responsibility to act to be based on contributions to the problem and capacity to respond 基于对问题的贡献和应对能力所采取行动的责任	2. Precautionary Principle 预防措施
C. Sustainable growth and development by all Parties, particularly developing countries 所有缔约方的可持续增长和发展，特别是发展中国家	3. Common but Differentiated Responsibilities 共同但有区别的责任

Source 资料来源：UNFCCC《联合国气候变化框架公约》。

Kyoto Protocol
《京都议定书》

The Kyoto Protocol is
《京都议定书》是

- An international treaty related to the UNFCCC
- Adopted at the Third Conference of the Parties held in Kyoto, Japan in 1997
- Became effective from February 2005
- There are currently 192 parties (as of 2020)
- The first commitment period is 2008—2012
- The second commitment period is 2013—2020

- 与UNFCCC相关的国际条约
- 1997年日本京都召开的第三次缔约方大会上通过
- 2005年2月生效
- 目前有192个缔约方（至2020年）
- 第一承诺期为2008—2012年
- 第二承诺期为2013—2020年

Significance of the Kyoto Protocol
《京都议定书》的重要性

Complement and strengthen the Convention
补充并加强《框架公约》

Clearly identified and specified the 6 GHGs
确立并规范了6种温室气体

Provide developing countries with opportunities to mitigate and adapt to climate change
为发展中国家提供减缓和适应气候变化的机会

Established binding emission limits for developed countries
为发达国家制定了单独的具有约束力的排放量

2008—2012年

First Commitment Period 第一承诺期

37 industrialized countries and the EU committed to:

- Reduce emissions of the 6 GHGs by at least 5% from 1990 levels over the 2008—2012 commitment period
- ensure that its CO_2 equivalent emission does not exceed its allocated amount

37个工业化国家和欧盟承诺：

- 在2008—2012年承诺期内这6种温室气体的总排放量至少较1990年减少5%
- 确保其二氧化碳排放当量不超过其分配数量

2013—2020年

Second Committment Period 第二承诺期

- Establishment of the *Doha Amendment*
- Reduce GHG emissions by at least 18% from 1990 levels
- The 37 countries participating in the second commitment period account for 14% of world emissions
- 《京都议定书》的《多哈修正案》
- 在1990年水平上至少减少18%的温室气体排放量
- 37个参与第二承诺期的国家的总排放量占全世界的14%

Read the information about *Kyoto Protocol* and the *Doha Amendment*, what are some improvements or progress in establishing climate agreements comparing to the 1992 version of the UNFCCC?

阅读《京都议定书》和《多哈修正案》相关资料，与1992年版的《框架公约》相比，气候协议的制定有哪些改进或进展？

Source 资料来源：UNFCCC《联合国气候变化框架公约》。

Answer Key 答案（P.31）： A. 2 B. 3 C. 1

Paris Agreement
《巴黎协定》

COP 21 adopted the Paris Agreement to achieve a world that is carbon-neutral and resilient to climate change by mid-century and keep the increase in global mean temperature to well below 2°C.

第二十一次缔约方大会上通过了《巴黎协定》，以在 21 世纪中叶实现一个碳中和并能抵御气候变化的世界为目标，并将全球平均气温的上升幅度控制在远低于 2℃的范围内。

Mechanisms of the Paris Agreement 《巴黎协定》的机制

Action Areas 行动领域

Global Temperature Goal-to keep the temperature increase to well below 2°C by 2100

Mitigation-to reduce GHG emissions reaching global peaking by 2050

Adaptation-is a goal for all Parties

全球气温目标——到 2100 年保持气温升高的幅度远低于 2℃
减缓——到 2050 年确保温室气体排放量达到全球峰值
适应——这是适用于所有缔约方的目标

Support Mechanisms 支援机制

Finance-developed countries to provide $100 bn/year from 2020

Capacity Building-to support the most vulnerable to mitigate and adapt

Technology-the Agreement will be supported by the UNFCCC Technology Mechanism and Framework

资金——发达国家从 2020 年起每年提供 1 000 亿美元
能力建设——支持最脆弱的群体实施缓解和适应措施
技术——该协议将得到 UNFCCC 技术机制和《框架公约》的支持

Review Mechanisms 审查机制

Transparency-to provide clarity on support provided and received by relevant individual Parties in the context of climate change actions

Global Stocktake-to measure progress on mitigation from 2023 and every five years from then

Compliance-to ensure that Parties meet their legally binding commitments

公开透明——明确相关个别缔约方在气候变化行动中提供和接受的支援
全球盘点——从 2023 年开始，此后每 5 年衡量缓解措施的进展情况
遵守——确保缔约方履行其具有法律约束力的承诺

Read the *Paris Agreement* and discuss the following questions:

What challenges do you see in achieving the goals of the Agreement?

From the current situation, do you think we are on track to achieve the goals set out in the Agreement?

What are your country's commitments or responsibilities under the *Paris Agreement*?

阅读《巴黎协定》，讨论以下问题：

你认为协议的达成面临着什么样的挑战？

从目前的情况来看，你觉得我们能达到《巴黎协定》中提出的目标吗？

你所在的国家在《巴黎协定》中的承诺或承担的责任是什么？

Source 资料来源：UNFCCC《联合国气候变化框架公约》。

UNSDGs-The United Nations Sustainable Development Goals
联合国可持续发展目标

At the three-day summit on sustainable development in 2015, more than 150 country leaders gathered at United Nations Headquarters in New York to formally approve an ambitious new agenda for sustainable development.

The new plan, called "Transforming Our World: The 2030 Agenda for Sustainable Development by 2030", included a declaration, 17 sustainable development goals and 169 targets. The goal of the plan: to find new ways to improve the lives of the world's people, to eradicate poverty, to promote prosperity and well-being for all, to protect the environment, and to fight against climate change.

2015 年，150 多位世界领导人齐聚纽约联合国总部召开了为期 3 天的可持续发展峰会，正式通过一项目标远大的可持续发展新议程。

该议程被称为“变革我们的世界：2030 年可持续发展议程”，其中包括 1 项宣言、17 个可持续发展目标和 169 个具体目标。新议程旨在寻找新的方式改善全世界人民的生活、消除贫困、增进所有人的健康与福祉、保护环境以及应对气候变化。

"Development that meets the needs of the present without compromising the ability of future generations to meet their own needs."

——Our Common Future, 1987

“可持续发展是指既能满足当代需要，而同时又不损及后代需求的发展模式。”

——《我们共同的未来》，1987

The 17 United Nations Sustainable Development Goals
17 个联合国可持续发展目标

Goal 1. End poverty in all its forms everywhere

目标 1：在全世界范围内消除一切形式的贫穷

Goal 2. End hunger, achieve food security and improved nutrition and promote sustainable agriculture

目标 2：消除饥饿，实现粮食安全，改善营养状况和促进可持续农业

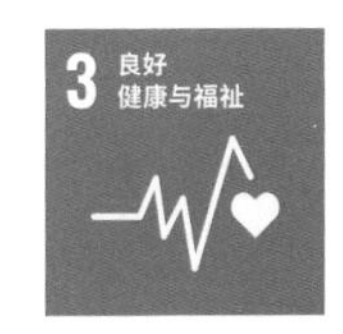

Goal 3. Ensure healthy lives and promote well-being for all at all ages

目标 3：确保健康的生活方式，促进各年龄段人群的福祉

Goal 4. Ensure inclusive and equitable quality education and promote lifelong learning opportunities for all

目标 4：确保包容和公平的优质教育，让全民终身享有学习机会

Goal 5. Achieve gender equality and empower all women and girls

目标 5：实现性别平等，增强所有妇女和女童的权利

Goal 6. Ensure availability and sustainable management of water and sanitation for all

目标 6：为所有人提供清洁饮水和卫生设施并对其进行可持续管理

UNSDGs-The United Nations Sustainable Development Goals
联合国可持续发展目标

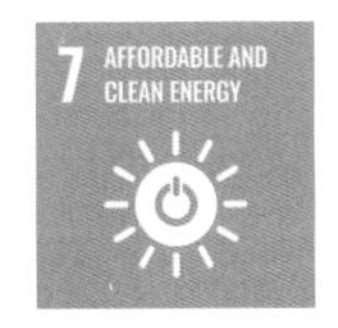

Goal 7. Ensure access to affordable, reliable, sustainable and modern energy for all
目标 7：确保人人获得可负担、可靠和可持续的现代能源

Goal 10. Reduce inequality within and among countries
目标 10：减少国家内部和国家之间的不平等

Goal 13. Take urgent action to combat climate change and its impacts
目标 13：采取紧急行动应对气候变化及其影响

Goal 16. Promote peaceful and inclusive societies for sustainable development, provide access to justice for all and build effective, accountable and inclusive institutions at all levels
目标 16：倡建和平、包容的社会以促进可持续发展，让所有人都能诉诸司法，并在各级建立有效、负责和包容的机构

Goal 8. Promote sustained, inclusive and sustainable economic growth, full and productive employment and decent work for all
目标 8：促进持久、包容和可持续的经济增长，促进充分的生产性就业并且人人获得体面工作

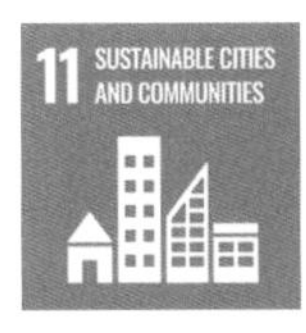

Goal 11. Make cities and human settlements inclusive, safe, resilient and sustainable
目标 11：建设包容、安全、有抵御灾害能力和可持续的城市和人类居住区

Goal 14. Conserve and sustainably use the oceans, seas and marine resources for sustainable development
目标 14：保护和可持续利用海洋和海洋资源以促进可持续发展

Goal 17. Strengthen the means of implementation and revitalize the global partnership for sustainable development
目标 17：加强执行手段，重振可持续发展全球伙伴关系

Goal 9. Build resilient infrastructure, promote inclusive and sustainable industrialization and foster innovation
目标 9：建设具有适应力的基础设施，促进包容性和可持续的工业化，推动创新

Goal 12. Ensure sustainable consumption and production patterns
目标 12：确保采用可持续的消费和生产模式

Goal 15. Protect, restore and promote sustainable use of terrestrial ecosystems, sustainably manage forests, combat desertification, and halt and reverse land degradation and halt biodiversity loss
目标 15：保护、恢复和促进可持续利用陆地生态系统，可持续管理森林，防治荒漠化，制止和扭转土地退化，遏制生物多样性的丧失

UNSDGs-The United Nations Sustainable Development Goals
联合国可持续发展目标

Reading the *2030 Agenda for Sustainable Development*, what are the five key themes considered? Please categorize the 17 SDGs under five themes. (see next page for answers)

阅读《2030 年可持续发展议程》，议程中提到了哪 5 个关键主题？请将 17 个可持续发展目标按照 5 个主题进行分类。（答案见下页）

联合国 A/RES/70/1

大 会

Distr.: General
21 October 2015

第七十届会议

2015 年 9 月 25 日大会决议

70/1. 变革我们的世界：2030 年可持续发展议程

大会

变革我们的世界：2030 年可持续发展议程

序言

Collaborative SDGs actions need to be taken at the individual level, the corporate level, national level and international level.

Could you find a SDG Action that has been implemented in your local community?

为达成可持续发展目标，需要从个人、企业、国家和国际层面协调同步展开行动。

你能在自己的社区中找出已经实施的可持续发展目标行动吗？

Another way of categorization
3 pillars
of the SDGs

另一种分类方法
可持续发展目标的 3 个支柱

Agreements on Major Topics
针对主要议题的协定

Classroom Activity: Profile Exchange
课堂活动：交换档案

To achieve climate targets, the United Nations has created numerous climate agreements on a wide range of topics, covering adaptation, mitigation, technology, finance, etc. Below are a few examples of climate agreements. Choose a climate agreement that you are interested in or that you think is the most important and the most effective to implement, do some research and make a "profile" for the agreement, you could consider the following features. Exchange it with a classmate, and discuss why you choose this agreement.

为了实现气候目标，联合国就广泛的主题制定了许多气候协议，涵盖适应、减缓、技术、金融等。以下是一些气候协议的例子。选择一份你感兴趣或你认为最重要、实施得最有效的协议，做些调查，为它制作一份“个人档案”，你可以考虑以下几个因素。与同学交换档案，讨论选择这份协定的原因。

Year of Establishment
建立年份

Review Mechanism
审查机制

Major Parties
主要参与方

Expected Outcome
预期效果

Implementation Strategies
实施策略

Main Aim
主要目标

Adaptation
适应

- Cancun Adaptation Framework
  《坎昆协议》(CAF)
- National Adaptation Plans
  《国家适应计划》(NAPs)
- National Adaptation Programme of Action
  《国家适应行动计划》(NAPAs)

Mitigation
减缓

- Reducing Emissions from Deforestation and Forest Degradation in Developing Countries
  减少发展中国家森林砍伐与森林退化所导致的碳排放 (REDD+)
- Nationally Appropriate Mitigation Action
  国家适用减缓计划 (NAMA)

Technology
技术

- Climate Technology Centre and Network
  气候技术中心和网络 (CTCN)
- Technology Mechanism
  技术机制

Finance
资金

- Fast-Start Finance
  快速启动资金
- Long-Term Climate Finance
  长期气候融资 (LTF)

Answer Key
答案（P.36）：
1. People 人类
2. Planet 地球
3. Prosperity 繁荣
4. Peace 和平
5. Partnership 伙伴关系

Speech Activities
演讲活动

Every climate change agreement or declaration in the past has been touching and ambitious. Each iconic document reflects humanity's awareness of the importance of sustainable development and our determination to take action. The documents urges all countries, businesses, NGOs, and individuals to work together to solve the challenges facing both humanity and the environment, reshape our relationship with the planet, and create a better future. Try to read an agreement carefully and pick the part that inspires you the most. Imagine you are announcing the results of the agreement in the Hall of the United Nations conference and give a touching speech in front of your class. Students can vote for the most inspiring speeches. The example below is an excerpt from the Stockholm Declaration.

以往的每一份气候变化协议或宣言都是触动人心、雄心勃勃的。每份标志性文件都体现了人类对可持续发展重要性的深刻认识和为之付诸行动的决心。这些文件呼吁所有国家、企业、非政府组织、个人共同合作，携手面对人类和环境共同面临的危机，重塑人与地球的关系，创造更美好的未来。仔细阅读一份协议并挑选其中最触动你的部分，想象你正在联合国会议的大礼堂内宣读协议结果，然后在全班同学面前进行演讲。同学们可投票选出最振奋人心的演讲。以下示例是《斯德哥尔摩宣言》的节选。

1. Man is both creature and moulder of his environment, which gives him physical sustenance and affords him the opportunity for intellectual, moral, social and spiritual growth. In the long and tortuous evolution of the human race on this planet a stage has been reached when, through the rapid acceleration of science and technology, man has acquired the power to transform his environment in countless ways and on an unprecedented scale. Both aspects of man's environment, the natural and the man-made, are essential to his well-being and to the enjoyment of basic human rights the right to life itself...

6. A point has been reached in history when we must shape our actions throughout the world with a more prudent care for their environmental consequences ... To defend and improve the human environment for present and future generations has become an imperative goal for mankind-a goal to be pursued together with, and in harmony with, the established and fundamental goals of peace and of worldwide economic and social development.

7. To achieve this environmental goal will demand the acceptance of responsibility by citizens and communities and by enterprises and institutions at every level, all sharing equitably in common efforts. Individuals in all walks of life as well as organizations in many fields, by their values and the sum of their actions, will shape the world environment of the future.

Source 资料来源：Stockholm Declaration《斯德哥尔摩宣言》。

1. 人类既是自身环境的创造物，又是自身环境的塑造者，环境为人类提供了赖以生存的物质，以及在智力、道德、社交和精神等方面的发展机会。生活在地球上的人类，在漫长而曲折的进化过程中，已经达到这样一个阶段，即由于科学技术的迅速发展，人类有了各种改造环境的方法。人类环境的两个方面，即自然环境和人造环境，对于人类享受幸福和基本权利甚至生存权利都不可或缺……

6. 现在已经到了这样一个历史时刻：我们在决定世界各地的行动时，必须更加审慎地考虑它们对环境产生的影响……为了这一代和将来的世世代代，保护和改善人类环境已成为我们紧迫的目标。这个目标将同争取和平、全世界的经济与社会发展这两个既定的基本目标共同协调和实现。

7. 为实现这一环境目标，公民和团体以及各级企业和机构需共同承担责任、共同努力。通过各界人士和不同领域的机构的信念和所采取的行动，塑造未来世界的环境。

CHAPTER 4
单元四

TAKING CLIMATE ACTIONS

采取气候行动

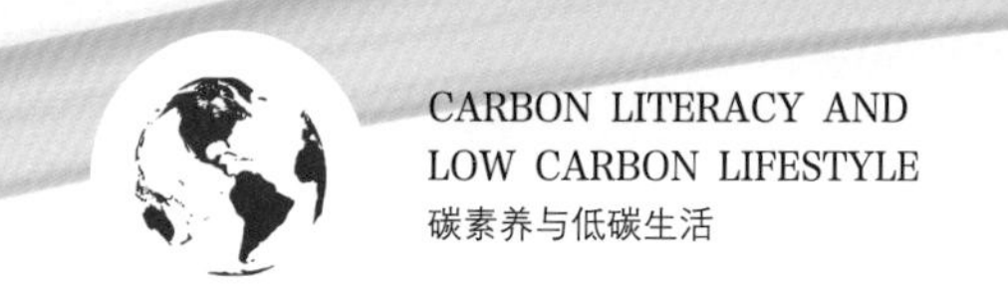

Impact of Our Activities on Climate Change
人类活动对气候变化的影响

7 AFFORDABLE AND CLEAN ENERGY 7 经济适用的清洁能源 9 INDUSTRY, INNOVATION AND INFRASTRUCTURE 9 产业、创新和基础设施

Power Generation/ 能源生产

Generating electricity and heat by burning fossil fuels causes a large chunk of global emissions. Most electricity is still generated by burning coal, oil, or gas, which produces carbon dioxide (CO_2) and nitrous oxide (N_2O), powerful greenhouse gases that blanket the Earth and trap the sun's heat. Globally, a bit more than a quarter of electricity comes from wind, solar and other renewable sources which, as opposed to fossil fuels, emit little to no greenhouse gases or pollutants into the air.

通过化石燃料燃烧发电和供热所产生的碳排放占据全球碳排放的大部分。大部分电力仍旧是通过燃烧煤炭、石油或天然气生产，其过程中会产生二氧化碳（CO_2）和一氧化二氮（N_2O）。这些强效的温室气体会覆盖地球表面并吸收太阳的热量。在全球范围内，仅超过 1/4 的电力来自风能、太阳能和其他可再生能源，与化石燃料相比，这些能源基本上不会向空气中排放温室气体或污染物。

Since 2000, the world has doubled its coal-fired power capacity from 1,066GW to around 2,045 gigawatts in 2019 (GW).

自 2000 年以来，全球燃煤发电量成倍增加，从 1 066 吉瓦上升到 2019 年的约 2 045 吉瓦。

According to the "Statistical Communiqué of the People's Republic of China on the 2022 National Economic and Social Development" issued by the National Bureau of Statistics of China, by the end of 2022, the total installed capacity of power generation in China was 2,564.05 million kilowatts, an increase of 7.8% comparing to 2021.

根据国家统计局发布的《中华人民共和国 2022 年国民经济和社会发展统计公报》，2022 年年末，全国发电装机容量 2 564.05 百万千瓦，较 2021 年年末增长 7.8%。其中：

Method of Electricity Generation 发电方式	Electricity Generation Capacity/million kW 发电装机容量 / 百万千瓦	Increase comparing to 2021/% 较 2021 年增长 /%	Amount of Electricity Generated/million kW·h 发电量 / 百万千瓦时	Increase comparing to 2021/% 较 2021 年增长 /%
Thermal power 火电	133 239	2.7	58 887.9	1.4
Hydropower 水电	41 350	5.8	13 522.0	1.0
Nuclear power 核电	5 553	4.3	4 177.8	2.5
Wind power 风电	36 544	11.2	7 626.7	16.2
Solar power 太阳能	39 261	28.1	4 272.7	31.2

Source 资料来源： National Bureau of Statistics of China 中华人民共和国国家统计局。

Impact of Our Activities on Climate Change
人类活动对气候变化的影响

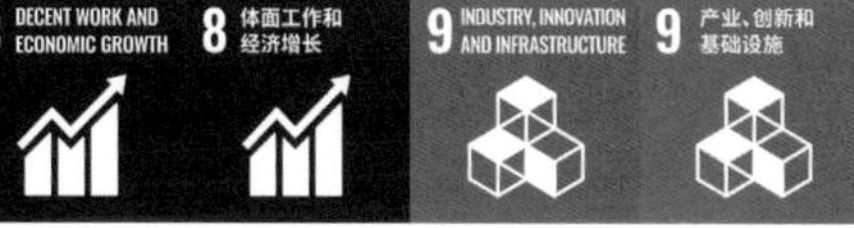

Manufacturing Goods/ 制造商品

Manufacturing and industry produce emissions, mostly from burning fossil fuels to produce energy for making things like cement, iron, steel, electronics, plastics, clothes, and other goods. Mining and other industrial processes also release gases, as does the construction industry. Machines used in the manufacturing process often run on coal, oil, or gas. Some materials, like plastics, are made from chemicals sourced from fossil fuels. Therefore, the manufacturing industry is one of the largest contributors to greenhouse gas emissions worldwide.

制造业和工业产生的温室气体排放主要来源于化石燃料燃烧，其目的是为制造水泥、钢铁、电子产品、塑料制品、服装和其他商品提供能源。采矿业和其他工业活动也会释放温室气体，建筑业也是如此。生产制造过程中使用的机器通常靠煤炭、石油或天然气供能运行。有些材料（如塑料）是由化石燃料中的化学物质制成的。因此，制造业是全球温室气体排放的最大来源之一。

Demand for steel, which drives steel production, is a key determinant of energy demand and steel subsector CO_2 emissions. Global crude steel production increased by an average of 3% per year. At the same time, the steel sector is still highly reliant on coal, which meets 75% of its energy demand. The direct CO_2 intensity of crude steel production has been relatively constant in the past few years, every ton of steel produced in 2018 emitted on average 1.85 tons of carbon dioxide, equating to about 8 percent of global carbon dioxide emissions.

钢铁需求的增加是钢铁产量激增，进而增加了能源需求和钢铁行业二氧化碳排放的关键因素。全球粗钢铁产量平均每年增长 3%，与此同时，钢铁行业仍高度依赖煤炭能源，煤炭满足了其 75% 的能源需求。过去几年粗钢铁生产所产生的二氧化碳直接排放强度基本没有变化，2018 年，每生产 1 吨钢铁平均排放 1.85 吨二氧化碳，大致相当于全球二氧化碳排放量的 8%。

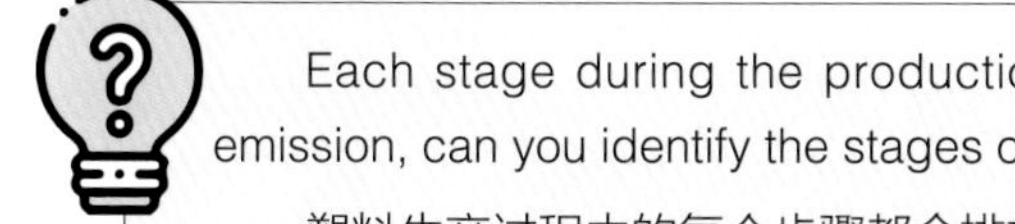

Each stage during the production of plastic contribute a significant amount of greenhouse gas emission, can you identify the stages of production illustrated below? (See next page for answers)

塑料生产过程中的每个步骤都会排放大量的温室气体，请识别下图所示的各个生产步骤。（答案见下页）

In the United States alone in 2015, emissions from fossil fuel A.______ and B.______ attributed to plastic production were at least 9.5～10.5 million metric tons of CO_2e per year. The C. ______ of plastic is both energy-intense and emissions-intensive, chemical refining processes emit a significant amount of GHG. Lastly, its waste processing is also carbon-intensive, global emissions from D.______ of plastic bag wastes totaled 16 million metric tons of CO_2e in 2015.

2015 年，美国塑料生产过程中的 A.______和 B.______环节所使用的化石燃料燃烧所产生的温室气体排放量至少有 9.5 万～10.5 万吨二氧化碳当量。塑料的 C.______消耗大量能源，同时它的化学精炼过程也会排放大量温室气体。最后，它的废物处理过程也是碳密集型的，2015 年，塑料袋废物 D.______所产生的全球排放总量为 1 600 万吨二氧化碳当量。

Source 资料来源：IEA 国际能源署；McKinsey & Company 麦肯锡咨询公司。

Impact of Our Activities on Climate Change
人类活动对气候变化的影响

Cutting Down Forests/ 砍伐树木

Forests are a stabilising force for the climate. They regulate ecosystems, protect biodiversity, play an integral part in the carbon cycle, support livelihoods, and supply goods and services that can drive sustainable growth.

Forests' role in climate change is two-fold. They act as both a cause and a solution for greenhouse gas emissions. Cutting down forests to create farms or pastures, or for other reasons, causes emissions, since trees, when they are cut, release the carbon they have been storing. Each year approximately 12 million hectares of forest are destroyed. This is why around 25% of global emissions come from the land sector, in which about half of these comes from deforestation and forest degradation.

Forests are also one of the most important solutions to addressing the effects of climate change. Since forests absorb carbon dioxide, destroying them also limits nature's ability to keep emissions out of the atmosphere. Approximately 2.6 billion tonnes of carbon dioxide, one-third of the CO_2 released from burning fossil fuels, is absorbed by forests every year. Estimates show that nearly two billion hectares of degraded land across the world, an area the size of South America, offer opportunities for restoration. Increasing and maintaining forests is therefore an essential solution to climate change.

森林是稳定气候的关键，它能够调节生态系统、保护生物多样性，在碳循环中发挥着不可或缺的作用，同时它也为人们带来收入并提供能够推动可持续经济增长的商品和服务。

森林在气候变化中有双重身份，它既是温室气体排放源也是气候变化的解决方案。砍伐森林以建立农场或牧场会排放温室气体，因为树木在被砍伐时会释放原本储存的二氧化碳。每年约有 1 200 万公顷的森林遭到破坏，全球约 25% 的排放来自土地相关部门，而其中大约一半来自森林砍伐和森林退化。

森林也是应对气候变化影响最重要的解决途径之一：森林能够吸收二氧化碳，因此破坏它们会限制大自然转化温室气体的能力。每年约有 26 亿吨二氧化碳被森林吸收，占燃烧化石燃料释放的二氧化碳的 1/3。然而，全球有近 20 亿公顷的森林土地已经退化，约相当于整个南美洲的面积！若我们努力恢复、维护这些森林，它们都能成为应对气候变化的解决途径。

Release CO_2 into the atmosphere
排放二氧化碳到空气中

Reduced ability to absorb CO_2
降低吸收二氧化碳的能力

CO_2 stored in trees
储存在树木中的二氧化碳

Source 资料来源：IUCN 世界自然保护联盟；UN 联合国。

Choose one of the forest from the list below, research about its current situation and discuss with your classmate: Is forest conservation effort more effective on the international level, the national level, or the local level?

从下面选择一处森林，搜集资料来了解它的现状，并与同学讨论：森林保护工作在国际层面、国家层面还是地方层面更有效？

- Amazon · 亚马孙
- Atlantic Forest · 大西洋森林
- Borneo · 婆罗洲
- Cerrado · 巴西热带稀树草原
- Eastern Australia · 澳大利亚东部
- Congo Basin · 刚果盆地

Answer Key 答案（P.41）： A. Extraction 开采 B. Transportation 运输 C. Manufacturing 制造 D. Incineration 焚烧

Impact of Our Activities on Climate Change
人类活动对气候变化的影响

Using Transportation/ 使用交通工具

Most cars, trucks, ships, and planes run on fossil fuels. That makes transportation a major contributor of greenhouse gases, especially carbon-dioxide emissions. Road vehicles account for the largest part, due to the combustion of petroleum-based products, like gasoline, in internal combustion engines. But emissions from ships and planes continue to grow. Transport accounts for nearly one quarter of global energy-related carbon-dioxide emissions. And trends point to a significant increase in energy use for transport over the coming years.

大部分汽车、货车、轮船和飞机都依靠化石燃料提供能源，这使得交通工具成为温室气体，尤其是二氧化碳排放的主要来源。由于公路汽车的内燃机燃烧的是汽油等以石油为基础的燃料，其排放量是最大的。同时，船舶和飞机的排放量也在持续增长。交通运输所产生的二氧化碳排放量约占全球能源相关碳排放量的 1/4。趋势表明，未来几年交通工具的能源消耗量还将继续大幅增加。

Distribution of CO_2 Emissions Produced by the Transportation Sector Worldwide
全球交通运输行业二氧化碳排放量分布

In 2019, global transportation is responsible for 24% of direct CO_2 emissions from fuel combustion. Road vehicles-cars, trucks, buses and two-and three-wheelers-account for nearly three-quarters of transport CO_2 emissions, and emissions from aviation and shipping continue to rise.

2019 年，全球交通运输所产生的二氧化碳排放量占燃料燃烧产生的直接二氧化碳排放量的 24%。公路汽车，包括汽车、货车、公共汽车以及两轮车和三轮车，占交通运输二氧化碳排放量的近 3/4。同时，航空和航运的二氧化碳排放量也在持续上升。

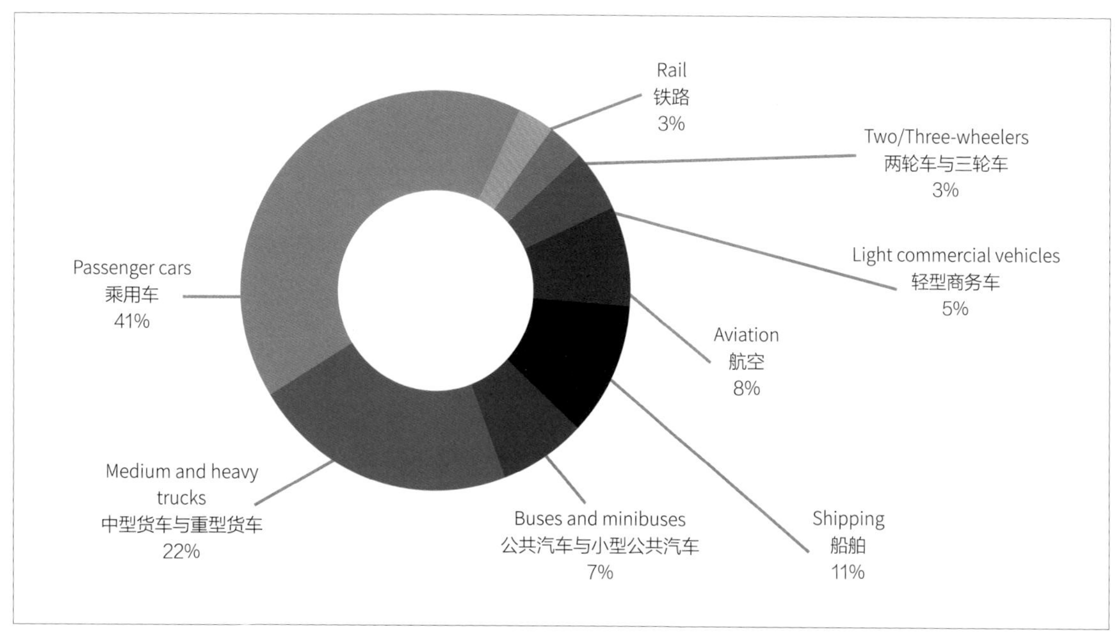

Distribution of CO_2 Emissions Produced by the Transportation Sector Worldwide, 2020
2020 年全球交通运输行业二氧化碳排放量分布

Source 资料来源：IEA 国际能源署。

Impact of Our Activities on Climate Change
人类活动对气候变化的影响

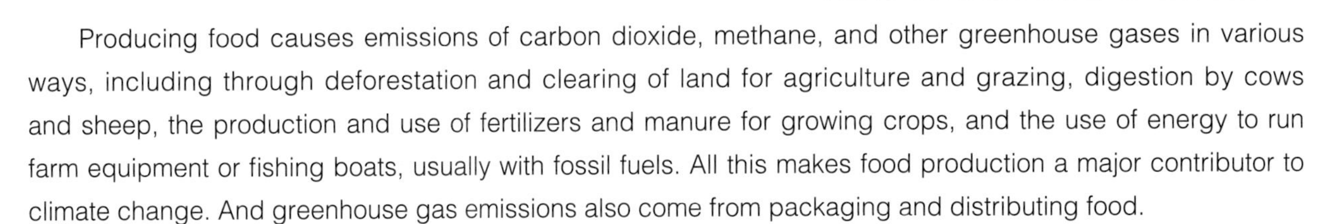

Producing Food/ 粮食生产

Producing food causes emissions of carbon dioxide, methane, and other greenhouse gases in various ways, including through deforestation and clearing of land for agriculture and grazing, digestion by cows and sheep, the production and use of fertilizers and manure for growing crops, and the use of energy to run farm equipment or fishing boats, usually with fossil fuels. All this makes food production a major contributor to climate change. And greenhouse gas emissions also come from packaging and distributing food.

粮食的生产过程中会以各种方式排放二氧化碳、甲烷和其他温室气体。这些方式包括为农耕和放牧而砍伐森林和开垦土地、牛羊消化食物时排放的气体、生产和使用化肥和粪肥来种植作物，以及使用化石燃料等能源来驱动农业设备或渔船。所有这些活动都使粮食生产成为导致气候变化的一个主要因素。此外，粮食的包装和分销过程也会排放温室气体。

Systems to produce, package and distribute food generate a third of greenhouse gas emissions and cause up to 80 per cent of biodiversity loss. Without intervention, food system greenhouse gases emissions will likely increase by up to 40 percent by 2050, given rising demand from population, more income and dietary changes.

食品生产、包装和分销系统产生的温室气体排放量占全球排放量的 1/3，并导致高达 80% 的生物多样性丧失。由于人口需求增加、收入提升和饮食结构的变化，如果不采取措施，到 2050 年粮食系统的温室气体排放量可能会增加多达 40%。

How does the food we eat contribute to global emission?
我们的饮食如何加剧全球排放？

Each stage in the production chain of the food we eat contribute different amount of emission, identify the sources of emission in each stage and possible emission reduction strategies.

我们所吃的食物在生产过程中的每个阶段都会产生不同的碳排放量。试找出每个阶段的主要排放源分别是什么，并提出可行的减排策略。

Driver 源头	Emission source 排放源	How can we reduce it 如何减排
Land Use Change 土地用途变化		
Farm 耕种		
Animal Feed 动物饲料		
Processing 食物处理		
Transport 运输		
Retail 销售		
Packaging 包装		

Source 资料来源：UN 联合国。

Impact of Our Activities on Climate Change
人类活动对气候变化的影响

Producing Food/ 粮食生产

Although food production causes significant GHG emissions, the problem of food waste has always been prominent. Food waste reduction offers multi-faceted wins for people and the planet, improving food security, addressing climate change, saving money and reducing pressures on land, water, biodiversity and waste management systems. Yet this potential has until now been woefully under-exploited.

The United Nations Sustainable Development Goal 12.3 (SDG 12.3) captures the commitment that: By 2030, halve per capita global food waste at the retail and consumer levels and reduce food losses along production and supply chains, including post-harvest losses.

尽管食品生产造成大量温室气体排放，但食物浪费问题一直都十分严重。减少食物浪费为人类和地球带来多方面的好处，它能改善食品安全、应对气候变化、节省开支，并减少对土地、水、生物多样性和废物管理系统的压力。然而，到目前为止，这一潜力仍未得到充分利用。

联合国可持续发展目标 12.3(SDG12.3) 承诺：到 2030 年，将零售和消费层面的全球人均粮食浪费量减半，同时减少生产和供应环节的粮食损失，包括收获后的损失。

Activity: Are you a food waster?
活动：你浪费了多少食物?

Track your food wastage for a week to get an idea of how much food you waste. We need to be more mindful of our food wastage!

记录你一周的食物浪费量。我们要更加注意避免食物浪费！

	Mon. 星期一	Tue. 星期二	Wed. 星期三	Thu. 星期四	Fri. 星期五	Sat. 星期六	Sun. 星期日
Breakfast 早餐							
Lunch 午餐							
Dinner 晚餐							
Snack 小食							

Source 资料来源：UNEP 联合国环境规划署。

Impact of Our Activities on Climate Change
人类活动对气候变化的影响

Powering Buildings/ 供能建筑

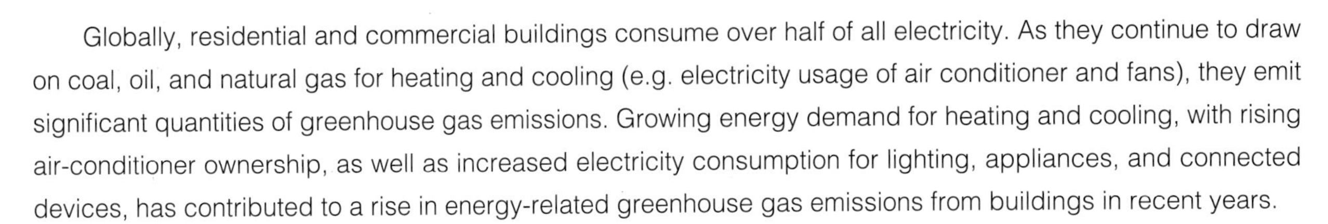

Globally, residential and commercial buildings consume over half of all electricity. As they continue to draw on coal, oil, and natural gas for heating and cooling (e.g. electricity usage of air conditioner and fans), they emit significant quantities of greenhouse gas emissions. Growing energy demand for heating and cooling, with rising air-conditioner ownership, as well as increased electricity consumption for lighting, appliances, and connected devices, has contributed to a rise in energy-related greenhouse gas emissions from buildings in recent years.

由于民用住宅和商业建筑消耗了全球一半以上的电力，同时这些建筑仍在使用煤炭、石油和天然气来供暖和制冷（空调、风扇等电器使用的电力），因此它们是不可忽视的温室气体排放源。近年来，随着空调的普及，供暖和制冷的能源需求不断增长，同时照明、电器和联网设备的用电量增加，与建筑物能源相关的温室气体排放量逐年上升。

Electricity consumption by household appliances continues to increase. It reached over 3,000 TW·h in 2019 and accounted for 15% of global final electricity demand, or one-quarter of electricity used in buildings.

家用电器的用电量持续增加，在 2019 年超过 3.0×10^{12} 千瓦时，占全球终端电力需求的 15%，也就是建筑物用电量的 1/4。

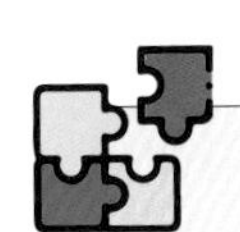

Activity: Identifying emission reduction opportunities in the kitchen
活动：找出厨房中的减排潜力

Try to circle which appliances in the picture below could be replaced with lower carbon emission options? Search online to find suitable alternatives.

下图中哪些电器可以被替换成碳排放量更低的电器？请试着圈出来，并通过上网搜集资料，为它们找到合适的替代品。

Source 资料来源：IEA 国际能源署；UN 联合国。

Impact of Our Activities on Climate Change
人类活动对气候变化的影响

Overconsumption/ 过度消费

Your home and use of power, how you move around, what you eat and how much you throw away all contribute to greenhouse gas emissions. So does the consumption of goods such as clothing, electronics, and plastics. A large chunk of global greenhouse gas emissions are linked to private households. Our lifestyles have a profound impact on our planet. The wealthiest bear the greatest responsibility: the richest 1 per cent of the global population combined account for more greenhouse gas emissions than the poorest 50 per cent.

居家用电、交通工具、食物和垃圾，都会排放温室气体，服装、电子产品和塑料制品也同样如此。全球大部分的温室气体排放都与个人家庭生活有关，我们的生活方式对地球有着深远的影响。最富有的人应承担最大的责任：全球最富有的 1% 人口的温室气体排放量大于全球最贫穷的 50% 人口的总排放量。

How much do our wardrobes cost to the environment?
我们的衣柜对环境造成了多大的影响?

According to figures from the United Nations Environment Programme (UNEP), it takes 3,781 liters of water to make a pair of jeans, from the production of the cotton to the delivery of the final product to the store. That equates to the emission of around 33.4 kilograms of carbon dioxide equivalent.

The fashion industry is responsible for 10% of annual global carbon emissions, more than all international flights and maritime shipping combined. At this pace, the fashion industry's greenhouse gas emissions will surge more than 50% by 2030. However, our clothing consumption has not stopped surging, the average person today buys 60% more clothing than in 2000, and less than 1% of used clothing is recycled into new garments.

根据联合国环境规划署的数据，一条牛仔裤从生产棉花到将最终产品运送到商店需要消耗 3 781 升水。这相当于排放约 33.4 千克二氧化碳当量。

时装业每年的碳排放量占全球总碳排放量的 10%，超过所有国际航班和海运的总和。按照目前这个趋势，到 2030 年，时装业的温室气体排放量将激增 50% 以上。然而，我们的服装消费量还在不停攀升，截至 2019 年，普通收入的消费者所买的衣服数量较 2000 年时多 60%，但被回收制成新衣物的旧衣物数量不足 1%。

Rank the following clothing materials from the highest CO_2e emission to the lowest.
将以下服装材料根据二氧化碳排放当量从高到低排序。(答案见下页)

A. Cotton
棉布

B. Linen
亚麻布

C. Silk
丝绸

D. Wool
羊毛

Source 资料来源：The World Bank 世界银行；UN 联合国。

Climate Change Adaptation
气候变化适应

Look at the picture on the left: which frog is preventing climate change, which one is mitigating and which one is adapting?

观察左图：哪只青蛙在逃避气候变化，哪只在减缓，哪只在适应？

In human systems, climate change adaptation is the process of adjustment to actual or expected climate and its effects, in order to moderate harm or exploit beneficial opportunities.

——IPCC, Special Report 1.5°C

在人类系统中，气候变化适应是指为减轻温室气体造成的损害或有效利用机会而适应实际或预期的气候及其影响的过程。

——IPCC，1.5℃特别报告

Adaptive Capacity

The ability of systems, institutions, humans and other organisms to adjust to potential damage, to take advantage of opportunities, or to respond to consequences.

适应能力

系统、机构、人类和其他生物在适应潜在损伤、利用机会或应对气候变化后果方面的能力。

Resilience

The capacity of interconnected social, economic and ecological systems to cope with a hazardous event, trend or disturbance, responding or reorganising in ways that maintain their essential function, identity and structure.

抵抗力

一个社会系统或生态系统在吸收各种干扰的同时保持相同的基本结构和功能方式的能力、自我组织的能力及适应压力和变化的能力。

Vulnerability

Vulnerability encompasses a variety of concepts and elements including sensitivity or susceptibility to harm and lack of capacity to cope and adapt.

脆弱性

脆弱性包含多种概念和要素，包括对伤害的敏感性或易感性，以及缺乏应对和适应的能力。

What are Climate Risks?
气候风险是什么？

Source 资料来源：IPCC 联合国政府间气候变化专门委员会。

Example:

Hazard–Change in precipitation pattern, including the amount, the location it occurs and the frequency.

Exposure–Is the rain occuring in places where we have human and natural assets, infrastructure, communities?

Vulnerability–Does the system impacted have the attribute to respond or bounce back?

例子：

危害——降水模式的变化，包括降水量、发生地点和频率。

暴露——降雨是否发生在人口密集、自然资产丰富、有基础设施或社区的地方？

脆弱性——受影响的地区是否具有应对或迅速恢复的能力？

Answer Key
答案（P.47）：1. D 3. C 2. A 4. B

Climate Change Adaptation
气候变化适应

Climate change, including rising temperatures, changes in precipitation patterns, melting of snow and ice, rising sea levels, and changes in the frequency and intensity of extreme weather events, will affect nearly all sectors of society and economy, therefore, every sector should implement effective adaptation strategies.

气候变化，包括气温上升、降水模式变化、冰雪消融、海平面上升、极端天气频率和强度的改变等，其影响将覆盖几乎所有的社会和经济层面，因此每个部门都应该实施有效的适应方案。

Water System
水资源系统

Water demand increases with rising temperature, while water supply depends on precipitation patterns and temperature, and water infrastructure is vulnerable to extreme weather.

需水量随着温度升高而增加，而供水量取决于降水模式和温度，水利基础设施也十分容易受到极端天气的影响。

Transportation
运输系统

While transport infrastructure is designed to withstand a particular range of weather conditions, climate change would expose them to weather outside historical design criteria.

虽然交通基础设施在设计时考虑到了需要承受的特殊气候条件，但气候变化会使它们暴露于以往设计标准未能预计到的天气中。

Energy
能源

Changes in temperature would affect energy demand as they are used to keep buildings warm in winter and cool in summer. It also affects energy supply through the cooling of thermal plants, through wind, solar and water resources for power, and through transmission infrastructure.

为了保持建筑物冬暖夏凉，能源需求会随着温度变化而改变。同时，热力发电厂的冷却需求，风能、太阳能和水资源发电，以及输电基础设施等也会影响能源供应。

Agriculture
农业

Agriculture is arguably the most climate-sensitive sector. A warming climate has a negative effect on crop production and generally reduces yields of staple cereals such as wheat, rice, and maize, which differ between regions and latitudes.

农业可以说是对气候最敏感的产业。气候变暖对农作物生产产生负面影响，这通常会降低小麦、稻米和玉米等主要谷物的产量，同时这些影响也因地区和纬度而异。

Activity: How will your region's agriculture be affected?
活动：你所在地区的农业会受到怎样的影响？

Research on the main agriculture production of your region, what are some of the weather conditions that determine their growth and yield? Compare with other regions, what are their unique vulnearbilities? You can consider the effect of temperature on crops, livestocks; how risks of flooding affect fisheries; forest fire as a threat to forestry.

调查你所在地区的主要农产品，哪些天气条件影响了它们的生长和产量？与其他地区相比，它们有哪些独特的脆弱性？思考一下气温对农业和畜牧业的影响、洪水风险对渔业的影响，以及森林火灾对林业构成的威胁。

Climate Change Adaptation
气候变化适应

Adaptation methods and cases/ 适应方法与案例

Risk Mitigation or Planned Adaptation
风险缓解或计划适应

This approach is applied when the climate risk is of high frequency and low or medium loss. For example, by planning for adaptation in the National Adaptation Programme of Action.

这个方法通常在气候风险频率较高、损失较低的情况下使用。例如，在国家适应行动计划当中制订适应计划。

Risk Transfer or Contingency Adaptation
风险转移或应急适应

This approach is applied when the climate risk is of low frequency and medium to high loss. Some extreme events can fall into this category, such as long-term droughts.

这个方法通常在气候风险频率较低、损失较高的情况下使用，例如应对长期干旱等极端事件。

Coping or Loss Acceptance
应对或接受损失

This approach is applied when the hazard is devastating but very unlikely to happen. Severe extreme events can fall into this category, such as unprecedented cyclones. Relief and humanitarian support measures are usually undertaken.

此方法通常在气候灾害达到毁灭性，但不太可能发生的情况下使用，例如前所未有的飓风。通常会采取救济和人道主义支援措施。

Discussion
讨论

What are some adaptation strategies that your country or region has implemented? Discuss with your classmate. Do you think they are effective?

你所在的国家或地区实施了哪些适应策略和措施？你认为它们有效吗？和同学讨论一下。

Source 资料来源：US EPA 美国国家环境保护局。

Vietnam 越南

Vietnam has made considerable progress in recent years to increase collaboration between state agencies responsible for disaster risk reduction and climate adaptation through the planning and implementation of joint initiatives. Each state agency keeps distinct mandates, roles and responsibilities, in accordance with Vietnam's National Adaptation Plan.

近年来，越南通过规划和实施联合倡议，在加强负责减少灾害风险和适应气候变化的国家机构之间的合作方面取得了显著的进展。根据越南的国家适应计划，每个国家机构都承担着不同的任务、角色和责任。

California 加利福尼亚州

Wildfires, a longstanding and frequent threat to California, are expected to increase in intensity and frequency due to climate change. Identified adaptation strategies in the 2018 California Climate Adaptation Strategy including fire suppression efforts, developing institutional capacity to monitor and mitigate the increased threat and risk of wildfires, increasing public awareness of proper land management strategies, and promoting efforts to better maintain air quality.

气候变化导致森林火灾频发是加利福尼亚州一直以来面临的威胁，其强度和频率预计将逐年增加。2018 年，加利福尼亚州制定了一系列气候适应战略，其中包括加强灭火工作、提升负责监测和减轻火灾危害的机构的能力、提高公众对适当的土地管理策略的认识，以及维持良好的空气质量等。

Climate Change Mitigation
气候变化减缓

Climate change mitigation is a human intervention to reduce emissions or enhance the sinks of greenhouse gases.

——IPCC, Climate Change 2022: Impacts, Adaptation and Vulnerability

减缓气候变化是指人类为减少碳排放或增加碳汇而采取的人为的干预措施。

——IPCC，气候变化 2022：影响、适应和脆弱性

1. Reduce GHGs emissions, such as improving the energy efficiency of older equipment
   例如可以提高旧设备的能源利用效率来减少温室气体排放
2. Prevent additional GHGs emissions from entering into the atmosphere, such as by avoiding building new emissions-intensive factories
   防止额外的温室气体排放到大气中，例如可以采取避免建设新的排放密集型工厂的方式
3. Maintain and enhance GHGs storage, such as protecting natural carbon sinks (forests,oceans, etc.) or creating new carbon sinks (carbon sequestration)
   保持并提高温室气体的储存，例如保护天然碳汇（森林、海洋等）或创造新的碳汇（碳封存）

Focusing on Emission Reduction in Key Sectors
聚焦关键部门的减排

Energy
能源

Renewable energy deployment requires redesigning our electricity markets, so that affordable and clean energy is delivered to consumers and industries. For example, mini-grids technologies allow for clean energy access in remote areas and are impacting the lives of millions.

可再生能源的部署需要我们重新规划电力市场，以便向消费者和相关行业提供负担得起的清洁能源。例如，微型电网技术能帮助偏远地区获得清洁能源，并为数百万人的生活带来积极影响。

Is nuclear energy a renewable energy?
核能是可再生能源吗?

Nuclear energy is a clean energy as nuclear power plants do not pollute the air or emit greenhouse gases. However, the material most often used in nuclear power plants is the element uranium, although uranium is found in rocks all over the world, nuclear power plants usually use a very rare type of uranium (U-235), which makes it a non-renewable energy source.

核能是清洁能源，因为核电站不会污染空气或排放温室气体。然而，核电站最常用的原材料是铀元素，虽然铀在世界各地的岩石中都有发现，但核电站通常使用一种非常稀有的铀（U-235），因此核能不是可再生能源。

Source 资料来源：National Geographic 国家地理。

What are the renewable energy illustrated in the pictures below? (See next page for answers)

下图中的设施分别代表哪些可再生能源？（答案见下页）

A.

B.

C.

D.

Climate Change Mitigation
气候变化减缓

Focusing on Emission Reduction in Key Sectors/ 聚焦关键部门的减排

Food
食物

1. Every year close to 1/3 of the food produced is wasted, generating 8% of global emissions! Utilizing food to the fullest is good for both the planet and people.
2. 1kg of beef meat generates 19 kg of CO_2e. 1kg of lentils generates less than 1kg of CO_2e. Plant-based alternatives to animal protein help consumers to reduce their carbon footprint.
3. Climate Smart Agricultural practices help farmers to sustainably increase productivity and system resilience while reducing greenhouse gas emissions.

1. 每年生产的食物中有近 1/3 被浪费，占全球 8% 的排放量。充分利用食物对地球和人类都是有利的。
2. 1 千克的牛肉会产生 19 千克二氧化碳当量的排放，而 1 千克扁豆只会产生少于 1 千克二氧化碳当量的排放。用植物蛋白代替动物蛋白可帮助消费者减少碳足迹。
3. 气候智能型农业生产方案的运用实施能够帮助农民在减少温室气体排放的同时持续提高农作物生产力和气候风险抵抗力。

Transportation
交通

1. Designing smart living spaces suitable for pedestrians and cyclists effectively reduces emissions from transportation and improves air quality.
2. In 2019, the share of global electric car sales was over 2%, up from 0.3% in 2014. Improved batteries and charging infrastructure will advance the shift.
3. A lot is also being done to reduce emissions through more efficient planes, fuel and flight-paths. In addition, the aviation sector has agreed to: (1) cap net emissions from 2020 onwards, (2) reduce net aviation emissions by 50% by 2050, relative to 2005.

1. 设计适合步行和单车出行的智能生活空间能有效减少交通排放，改善空气质量。
2. 2019 年，全球电动汽车销量占全球汽车总销量的比例超过 2%，远高于 2014 年总销量的 0.3%。电池和充电基础设施的进步将推进这一转变。
3. 航空业也正努力通过研发更高效的飞机、燃料，设计更短的飞行线路减少排放。此外，他们同意：（1）从 2020 年起限制航空业净排放量；（2）到 2050 年，将航空净排放量较 2005 年减少 50%。

Case: Electric Vehicle in Norway
案例：挪威普及电动汽车

The Norwegian Parliament has decided on a national goal that all new cars sold by 2025 should be zero-emission (electric or hydrogen). Battery electric vehicles held a 64% of its market share in 2021. Norway has been implementing EV Incentives since 1990. From 1996 to 2021, electric vehicles do not need to pay annual road tax. In 2001, it implemented a 25% purchase tax exemption policy. Many incentives helped Norway to approach a net-zero future.

挪威议会设定了一个国家目标，即确保到 2025 年销售的所有新车都是零排放的（电动或氢气）。电池电动汽车在 2021 年占据了其汽车市场 64% 的份额。挪威自 1990 年以来一直在实施电动汽车激励措施。1996—2021 年，电动汽车不需要缴纳年度道路税。2001 年实施 25% 的购置税免税政策。许多激励措施帮助挪威实现了接近净零的未来。

Source 资料来源：The One UN Climate Change Learning Partnership (UN:CC Learn) 联合国气候变化学习伙伴关系；Norsk elbilforening 挪威埃尔比福林。

Answer Key 答案（P.51）： A. Solar Energy 太阳能 B. Wind Energy 风能 C. Hydropower 水电 D. Geothermal Energy 地热能

Climate Change Mitigation
气候变化减缓

Increasing Carbon Sinks/ 增加碳汇

What are Carbon Sinks?
碳汇（直译为“二氧化碳下水槽”）是什么?

Carbon sinks are any process, activity or mechanism which removes a greenhouse gas, an aerosol or a precursor of a greenhouse gas from the atmosphere.

碳汇是指任何从大气中去除温室气体、气溶胶或温室气体前体的过程、活动或机制。

There are many approaches to reducing emissions and increasing carbon sinks. Some are old but effective methods, such as planting trees, while others rely on new, cutting edge technology. Categorise the following processes and activities into "A: reducing emission" or "B: increasing carbon sinks". Discuss with your classmates, which ones do you think have greater potential? Reflect on what the barriers may be in applying any of those solutions in your country. (See next page for answers)

有许多减少排放和增加碳汇的方法。一些是使用已久但有效的方法，例如植树，而另一些则依赖新的尖端科技。将以下过程和活动分类为 “A：减少排放” 或 “B：增加碳汇”。尝试与同学展开讨论，看看哪些有较大潜力。考虑你所在的城市或地区在应用这些解决方案时可能面临的挑战。(答案见下页)

1. Buying local products that have not travelled long distances
   购买没有经过长途货运的本地产品
2. Bio-energy power plants with carbon capture and storage
   具有碳捕获和储存功能的生物能源发电厂

3. Increase carbon content in soil
   增加土壤中的碳含量
4. Insulating houses
   隔热房屋
5. Efficient planes and increased use of bio-jet-fuel
   高效的飞机，增加生物喷气燃料的使用
6. Enhanced weathering of rocks (crushed rocks react with carbon in the air)
   加强岩石风化（碎石与空气中的碳发生反应）
7. Electric cars charged with renewable energy
   由可再生能源充电的电动车
8. No-tillage agriculture
   免耕
9. Use geothermal energy for heating
   使用地热能源供暖
10. Up-cycling, reusing and recycling products
    升级再造、再利用和回收产品
11. Reduce meat consumption
    减少肉类食用
12. Energy efficient lighting and home appliances
    节能照明和家用电器
13. Afforestation and Reforestation
    植树造林、恢复森林

Source 资料来源：The One UN Climate Change Learning Partnership (UN:CC Learn) 联合国气候变化学习伙伴关系；IPCC 联合国政府间气候变化专门委员会。

Climate Change Mitigation
气候变化减缓

Global Climate Change Mitigation Efforts/ 全球采取的气候变化减缓行动

The Race to Net Zero
达至净零的比赛

Put simply, net zero means cutting greenhouse gas emissions to as close to zero as possible, with any remaining emissions re-absorbed from the atmosphere, by oceans and forests for instance.

简单来说，净零意味着将温室气体排放量尽可能减少到接近于零，任何剩余的排放量都能从大气中被重新吸收，例如被海洋和森林吸收。

Transitioning to a net-zero world is one of the greatest challenges humankind has faced. It calls for a complete transformation of how we produce, consume, and move about. The energy sector is the source of around three-quarters of greenhouse gas emissions today and holds the key to averting the worst effects of climate change. Replacing polluting coal, gas and oil-fired power with energy from renewable sources, such as wind or solar, would dramatically reduce carbon emissions.

向净零的世界过渡是人类面临的最大挑战之一。我们必须彻底改变我们的生产、消费和行动方式。能源行业是当今约 3/4 温室气体的排放来源，是避免气候变化继续恶化的关键。以风能或太阳能等可再生能源取代污染严重的煤炭、天然气和石油发电，将大大减少碳排放。

Most emissions come from just a few countries.

Contribution of the 100 least-emitting countries is only 3%, whereas the 10 largest greenhouse gas emitters contribute over two-thirds of global emissions, approximately 68%.

大多数排放仅来自少数几个国家。

温室气体排放量最少的 100 个国家（地区）的排放量仅占全球排放总量的 3%，而温室气体排放量最多的 10 个国家（地区）的排放量占全球排放总量的 2/3 以上，约为 68%。

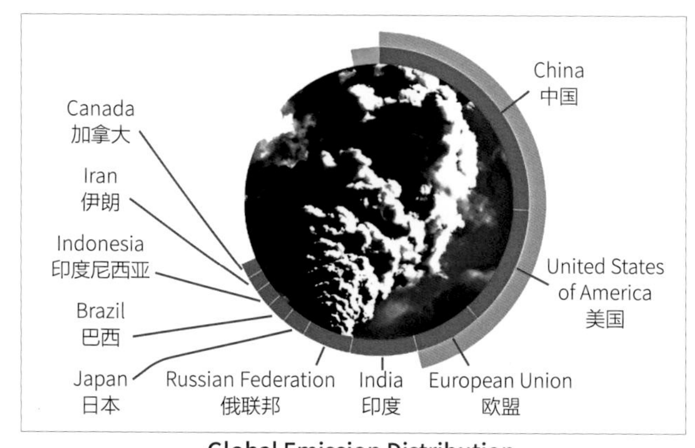

Global Emission Distribution
全球排放分布

A growing coalition of countries, cities, businesses and other institutions are pledging to get to net-zero emissions. More than 70 countries, including the biggest polluters–China, the United States, and the European Union–have set a net-zero target, covering about 76% of global emissions. Over 1,200 companies have put in place science-based targets in line with net zero, and more than 1,000 cities, over 1,000 educational institutions, and over 400 financial institutions have joined the Race to Zero, pledging to take rigorous, immediate action to halve global emissions by 2030.

越来越多的国家、城市、企业和机构正在承诺实现净零。70 多个国家（地区），包括作为“碳排放巨头”的中国、美国和欧盟都纷纷制定了净零目标，覆盖了全球排放量的约 76%。1 200 多家企业制定了以净零为基础的科学目标，1 000 多个城市、1 000 多家教育机构和 400 多家金融机构都加入了“达至净零的比赛”，承诺立即采取行动，目标是到 2030 年将全球排放量减少一半。

Source 资料来源：UN 联合国。

Answer Key 答案（P.53）：A, B, B, A, A, B, A, A, A, A, A, A, B

Climate Change Mitigation
气候变化减缓

Global Climate Change Mitigation Efforts/ 全球采取的气候变化减缓行动

Road to Net Zero
迈向净零之路

2015年

More than 190 countries adopted the historic *Paris Agreement* to reduce global warming and build resilience to climate change. Its overall goal: limit warming to no more than 1.5 degrees Celsius.

190 多个国家通过了具有历史意义的《巴黎协定》以减缓气候变暖并加强应对气候变化的能力。《巴黎协定》的整体目标是将全球升温控制在 1.5℃之内。

2015—2017年

Parties to the agreement began submitting climate action plans known as nationally determined contributions (NDCs). Initial commitments, even if fully implemented, would only be enough to slow warming to 3 degrees. Urgent calls for action and ambition gained momentum as the plans would not stop catastrophic impacts.

达成协定的各方开始提交气候行动计划，即国家自主贡献。然而这些初步承诺即使全部履行，也仅能将升温幅度放缓至 3℃。

2020—2021年

In the lead-up to the COP26 climate talks, countries have begun revising their NDCs to strengthen climate action. With science affirming a shrinking window of opportunity, the plans must include urgent actions to cut carbon emissions and reach net zero by 2050.

在联合国气候变化框架公约第 26 次缔约方大会气候谈判（COP26）之前，各国开始修订其国家自主贡献以加强气候行动。随着气候变化科学研究越发成熟，各项计划必须转化为迫切的行动以减少碳排放，从而实现 2050 年净零排放的目标。

2030年

To keep warming to 1.5 degrees, countries must cut emissions by at least 45 per cent compared to 2010 levels.

为了将全球升温控制在 1.5℃以内，各国必须在 2010 年的水平上至少将排放量减少 45%。

2050年

The transition to net zero must be fully complete.

全球必须完全转变为净零排放。

China
中国

China will strive to peak carbon dioxide emissions before 2030 and achieve carbon neutrality before 2060.

中国将力争在 2030 年前实现碳达峰、2060 年前实现碳中和。

India
印度

India will meet a target of net zero emissions by 2070.

印度将在 2070 年前实现净零排放目标。

United Kingdom
英国

UK aims to reach net zero emissions by 2050.

英国的目标是于 2050 年前实现净零排放目标。

Finland
芬兰

Finland aims to be carbon-neutral and the first fossil-free welfare society by 2035.

芬兰的目标是于 2035 年前达到碳中和，并成为第一个无化石补贴的国家。

Source 资料来源：UN 联合国。

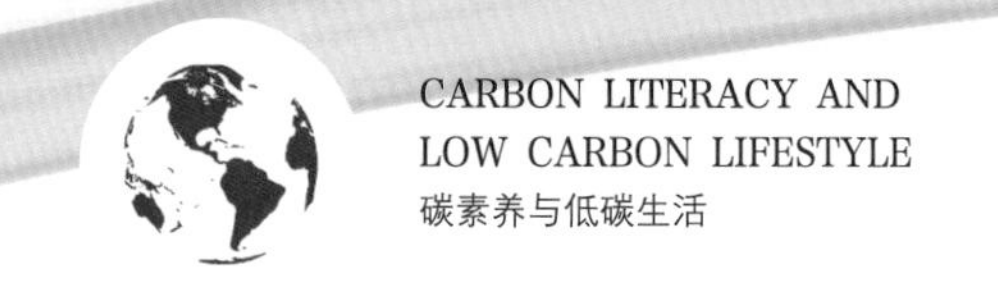

Climate Change Mitigation
气候变化减缓

Global Climate Change Mitigation Efforts/ 全球采取的气候变化减缓行动

Who has already achieved net zero?
谁已经实现了净零?

Bhutan is not only a popular travel destination, it is also the first country to become "carbon negative", with a bold promise to remain carbon neutral for all time. According to its report to UNFCCC in 2015, the estimated sequestration capacity of their forest is 6.3 million tons of CO_2 while the emissions for year 2000 is only 1.6 million tons of CO_2 equivalent. This is largely due to huge areas of forest cover, low levels of industrial activity and almost 100% electricity generation through hydropower.

不丹不仅是一个旅游胜地，还是全球首个达成“负碳”的国家，并承诺永远维持碳中和。根据不丹在 2015 年向 UNFCCC 提交的报告，其森林的碳封存 / 碳汇能力为大约 630 万吨二氧化碳，而不丹在 2000 年的排放量仅为160万吨二氧化碳当量。这主要是由于当地有大面积的森林覆盖、低工业活动水平以及近乎100% 的水力发电。

The other country that has reached net zero is Suriname, it stated as far back as 2014 that it had a carbon negative economy. Research on what policies and practices has Suriname implemented to pave its way towards a carbon negative economy, is there anything that could also be implemented in your country or region? What can we learn from Suriname?

另一个达到碳中和排放的国家是苏里南，早在 2014 年苏里南就表示已经达成负碳经济。通过资料搜集了解苏里南为建设负碳经济实施了哪些政策和措施？其中有哪些是你所在的国家或地区可以借鉴的？我们可以从苏里南的例子中学到什么？

Are we on track to reach net zero by 2050?
我们能达到 2050 年实现净零排放的预期目标吗?

According to the UN, as of 2021, the national climate plans–for all 190+ Parties to the *Paris Agreement* taken together–would lead to a sizable increase of almost 14% in global greenhouse gas emissions by 2030, compared to 2010 levels. Whereas we are aiming for a 45% cut in emissions compared to 2010 levels by 2030. This means the current effort is not sufficient, getting to net zero requires all governments to significantly strengthen their Nationally Determined Contributions (NDCs) and take bold, immediate steps towards reducing emissions now.

根据联合国数据，截至 2021 年，《巴黎协定》的 190 多个缔约方将令 2030 年全球温室气体排放量较 2010 年水平大幅增加近 14%。然而，我们的目标是在 2030 年将排放量与 2010 年的水平相比减少 45%。这意味着目前的努力还远远不够，实现净零排放需要所有政府大幅加强其“国家自主贡献”，迅速并果断地采取缩减排放的行动。

Do you think the net zero ambition achievable?
你认为净零排放的目标可以如期实现吗?

Source 资料来源：UN 联合国。

China's Climate Action
中国的气候行动

China's Climate Committements/ 中国的气候承诺

1998年

In 1998, China signed the *Kyoto Protocol* and ratified it in 2002.

中国于1998年签署《京都议定书》并于2002年批准。

2016年

China signed the *Paris Agreement* on April 22, 2016, on the first day it was opened for signature, and ratified it on 3 September, demonstrating leadership in G20 to issue the first presidential statement on climate change and provided political support for the signing of the *Paris Agreement*.

2016年4月22日是《巴黎协定》开放签署的首日，中国当日便签署了协定，并于9月3日批准协定。作为主席国，中国在推动二十国集团中发挥领导作用，首次发表关于气候变化的主席声明，为推动《巴黎协定》的签署提供政治支持。

2020年

On September 22, 2020, President Xi Jinping stated at the General Debate of the 75th Session of The United Nations General Assembly that "China will scale up its Intended Nationally Determined Contributions by adopting more vigorous policies and measures. We aim to have CO_2 emissions peak before 2030 and achieve carbon neutrality before 2060."

2020年9月22日，国家主席习近平在第七十五届联合国大会一般性辩论上表示，“中国将提高国家自主贡献力度，采取更加有力的政策和措施，二氧化碳排放力争于2030年前达到峰值，努力争取2060年前实现碳中和。”

2021年

On September 21, 2021, President Xi Jinping attended the General Debate of the 76th Session of The United Nations General Assembly, stating that "China will step up support for other developing countries in developing green and low-carbon energy, and will not build new coal-fired power projects abroad."

2021年9月21日，国家主席习近平在第七十六届联合国大会一般性辩论上表示，“中国将大力支持发展中国家能源绿色低碳发展，不再新建境外煤电项目。”

2021年

On October 24, 2021, the *Working guidance for carbon dioxide peaking and carbon neutrality in full and faithful implementation of the new development philosophy* was issued. As the "1" in the carbon neutrality "1+N" policy system to reach Carbon Peak, the document aims to carry out systematic planning and comprehensive deployment.

2021年10月24日，中共中央及国务院印发《关于完整准确全面贯彻新发展理念做好碳达峰碳中和工作的意见》。作为“1+N”政策体系中的“1”，为碳达峰及碳中和进行有系统的谋划及全面部署。

Do you know what is the "N" in the "1+N" policy system refers to? It could be found in the *Action Plan for Peaking Carbon Emissions Before 2030*. Try to list the ten key tasks included in "N" through online research.

你知道“1+N”政策体系中的“N”是指什么吗？
请参考《2030年前碳达峰行动方案》，并通过上网搜集资料，列出“N”中包含的十项重点任务。

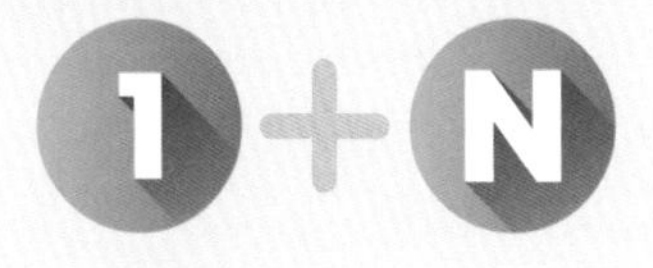

Source 资料来源：Full Text of President Xi Jinping's Statement at the General Debate of the 76th Session of the UNGA
国家主席习近平在第七十六届联合国大会一般性辩论上的讲话。

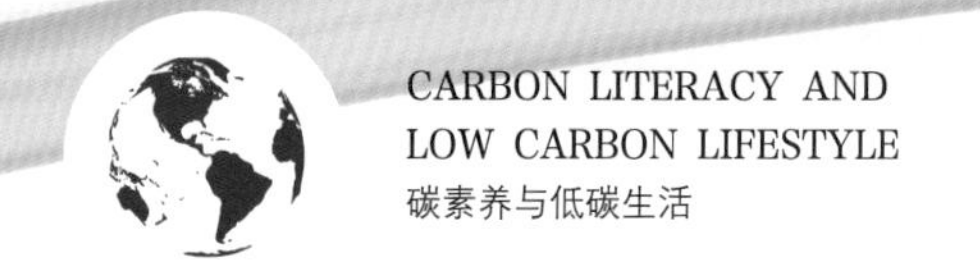

China's Climate Action
中国的气候行动

"N" in the *Action Plan for Peaking Carbon Emissions Before 2030*
《2030 年前碳达峰行动方案》中的“N”

Actions for green and low-carbon energy transition
能源绿色低碳转型行动

Vigorously promoting substitution of renewable sources of energy under the condition that energy security is ensured, and accelerate the development of a clean, low-carbon, safe and efficient energy system.

在保障能源安全的前提下，大力实施可再生能源替代，加快构建清洁低碳、安全高效的能源体系。

Developing hydro power according to local conditions
因地制宜开发水电

Promoting coal substitution as well as transformation and upgrading
推进煤炭消费替代和转型升级

Actively developing nuclear power through a safe and orderly approach
积极安全有序地发展核电

Speeding up the development of the new electric power system
加快建设新型电力系统

Vigorously developing new energy resources
大力发展新能源

Rationally regulating oil and gas consumption
合理调控油气消费

Actions for energy saving, carbon emission mitigation and efficiency improvement
节能降碳增效行动

Comprehensively improve energy-saving management capabilities, implement energy budget management, conduct comprehensive evaluations on energy consumption and carbon emissions of projects.

全面提升节能管理能力，推行用能预算管理，对项目用能和碳排放情况进行综合评价。

Implement urban energy-saving and carbon-reduction projects, carry out energy-saving upgrading and transformation of infrastructure such as buildings, transportation, lighting and heating.

实施城市节能降碳工程，开展建筑、交通、照明、供热等基础设施节能升级改造。

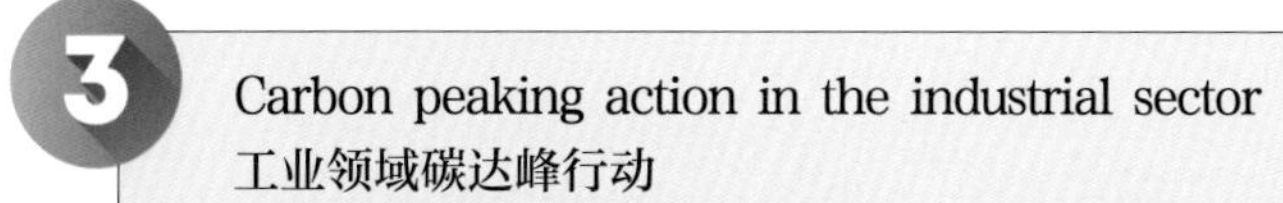

Carbon peaking action in the industrial sector
工业领域碳达峰行动

Promote the integrated development of digital, intelligent and green in the industrial field, and strengthen the technological transformation of key industries and fields.

推进工业领域数字化智能化绿色化融合发展，加强重点行业和领域的技术改造。

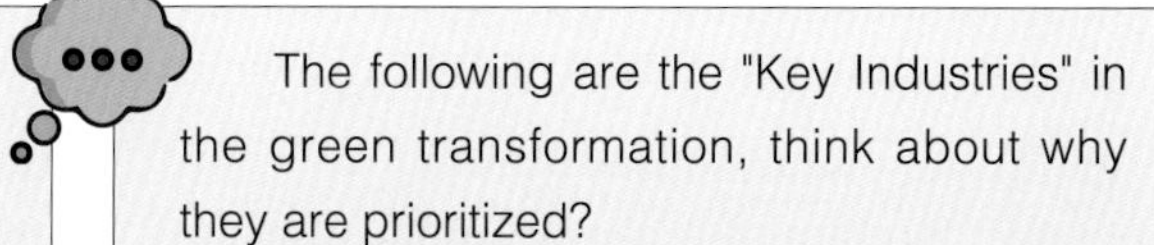

The following are the "Key Industries" in the green transformation, think about why they are prioritized?

试想想为什么我们要优先考虑绿色转型中的下列“重点行业”？

- Building materials 建材
- Non-ferrous metal 有色金属
- Petrochemical and chemical 石化化工
- Steel 钢铁

Source 资料来源：Circular of the State Council on *an action plan for peaking carbon emissions before 2030*
国务院关于印发《2030 年前碳达峰行动方案》的通知。

China's Climate Action
中国的气候行动

"N" in the *Action Plan for Peaking Carbon Emissions Before 2030*
《2030 年前碳达峰行动方案》中的"N"

Carbon peaking action in urban and rural construction
城乡建设碳达峰行动

- Advocate green and low-carbon planning and design concepts, promote the recycling of building materials, and strengthen green design and green construction management
- Accelerate the improvement of building energy efficiency
- Strengthen the research and development and promotion of energy-saving and low-carbon technologies suitable for different climate zones and different building types, and promote the large-scale development of ultra-low energy consumption buildings and low-carbon buildings
- 推广绿色低碳建材和绿色建造方式，推动建材循环利用，强化绿色设计和绿色施工管理
- 加快提升建筑能效水平
- 加强适用于不同气候区、不同建筑类型的节能低碳技术研发和推广，推动超低能耗建筑、低碳建筑规模化发展

Sun-Moon Altar - Minimal Carbon Building
日月坛·微排大厦

This building is the largest solar powered building in the world, with an area of 75,000 square meters, realizing the combination of solar hot water supply, heating, cooling, photovoltaic grid-connected power generation and other technologies. The energy efficiency reaches 88%, saving 2,640 tons of standard coal, 6.6 million kW·h of electricity and 8,672.4 tons of emissions every year!

日月坛·微排大厦是世界上最大的太阳能建筑，面积达 7.5 万平方米，在全球首创性地实现了太阳能热水供应、采暖、制冷、光伏发电等技术与建筑的结合，建筑整体节能效率达 88%，每年可节约标准煤 2 640 吨、节电 660 万千瓦时，减少污染物排放 8 672.4 吨。

In the *2020 China Green City Index TOP50 Report*, among 169 cities in China, Xiamen won the championship with a green index of 94.8, Shenzhen and Zhoushan ranked second and third with a score of 93.6 and 93.3 respectively. Have you been to any of these cities? Do you think they fulfill the title of "green city"?

《2020 中国绿色城市指数 TOP50 报告》显示，在生态环境部重点监测的 169 个城市当中，厦门以 94.8 的绿色指数夺冠，而深圳和舟山分别以 93.6 和 93.3 的得分排名第 2 位和第 3 位。你去过这些城市吗？你认为它们展现了"绿色城市"的风采吗？

Green and low-carbon actions in transportation
交通运输绿色低碳行动

Expand the application of new and clean energy such as electricity, hydrogen energy, natural gas, and advanced biological liquid fuels in the field of transportation.

扩大电力、氢能、天然气、先进生物液体燃料等新能源、清洁能源在交通运输领域的应用。

Accelerate the construction of urban and rural logistics distribution systems, and innovate green, low-carbon, intensive and efficient distribution models.

加快城乡物流配送体系建设，创新绿色低碳、集约高效的配送模式。

Create an efficient, fast and comfortable public transportation service system, and actively guide the public to choose green and low-carbon transportation methods.

打造高效衔接、快捷舒适的公共交通服务体系，积极引导公众选择绿色低碳交通方式。

Source 资料来源：Circular of the State Council on an *action plan for peaking carbon emissions before 2030*
国务院关于印发《2030 年前碳达峰行动方案》的通知。

China's Climate Action
中国的气候行动

"N" in the *Action Plan for Peaking Carbon Emissions Before 2030*
《2030 年前碳达峰行动方案》中的“N”

Circular economy fostering carbon reduction actions
循环经济助力降碳行动

- Improve resource output rate and recycling utilization rate
- Improve the recycling network of waste and used materials
- By 2025, the recycling volume of nine major renewable resources, including scrap iron and steel, scrap copper, scrap aluminum, scrap lead, scrap zinc, scrap paper, scrap plastic, scrap rubber, and scrap glass, reach 450 million tons, and reach 510 million tons by 2030
- Promote the reduction and recycling of domestic waste
- 提升资源产出率和循环利用率
- 完善废旧物资回收网络
- 到 2025 年，废钢铁、废铜、废铝、废铅、废锌、废纸、废塑料、废橡胶、废玻璃 9 种主要再生资源循环利用量达到 4.5 亿吨；到 2030 年达到 5.1 亿吨
- 全面实现生活垃圾分类投放、分类收集、分类运输、分类处理

Plastic Free policies
禁止塑料政策

Many cities in China have rigidly implemented policies to ban plastic:
中国多个地区也已实施严格的“禁塑”政策：

- From January 1, 2015, Jilin Province will prohibit the production, sale and provision of disposable non-degradable plastic film bags and tableware within the province's administrative region. Jilin Province has become the first province in the country to completely "ban plastics".
- In 2020, the use of plastic bags in shopping malls, supermarkets and other places in Zhejiang will be banned.
- From December 1, 2020, Hainan will completely "ban plastics".
- In April 2021, Beijing launched a "plastic ban" for food delivery, stipulating that non-degradable plastic bags and disposable plastic tableware are prohibited from being used in food delivery.

- 从 2015 年 1 月 1 日起，吉林省在全省范围内禁止生产、销售及提供一次性不可降解塑料购物袋和塑料餐具，成为全国首个全面“禁塑”的省级行政区。
- 2020 年，浙江省商场、超市等场所禁止使用塑料袋。
- 从 2020 年 12 月 1 日起，海南省全面“禁塑”。
- 2021 年 4 月，北京市推出餐饮外卖“禁塑令”，规定外卖禁止使用不可降解塑料袋和一次性塑料餐具。

Green and low-carbon technological innovation
绿色低碳科技创新行动

Improve innovative capacity building and talent training, incorporate green and low-carbon technological innovation achievements into relevant performance assessments of colleges and universities, scientific research units, and state-owned enterprises.

加强创新能力建设和人才培养，将绿色低碳技术创新成果纳入高等学校、科研单位、国有企业有关绩效考核。

Strengthen the main role of enterprises in innovation, support enterprises to undertake major national green and low-carbon science and technology projects, and encourage the open sharing of resources such as facilities and data.

支持企业承担国家绿色低碳重大科技项目，鼓励设施、数据等资源开放共享。

Source 资料来源：Circular of the State Council on an *action plan for peaking carbon emissions before 2030*
国务院关于印发《2030 年前碳达峰行动方案》的通知。

China's Climate Action
中国的气候行动

"N" in the *Action Plan for Peaking Carbon Emissions Before 2030*
《2030 年前碳达峰行动方案》中的“N”

8 Actions to consolidate and enhance carbon sink capacity
碳汇能力巩固提升行动

Strengthen the protection of forest resources and improve the quality and stability of forests
强化森林资源保护，
提高森林质量和稳定性

protect and restore marine ecosystems in a holistic way and improve the carbon sequestration capacity of mangroves, seagrass beds, and salt marshes
推进海洋生态系统保护和修复，提升
红树林、海草床、盐沼等固碳能力

Strengthen the restoration and management of degraded land
加强退化土地
修复治理

Strengthen grassland ecological protection and restoration
加强草原生态保护修复

Strengthen the protection and restoration of rivers, lakes and wetlands
加强河湖、湿地保护修复

- According to the 2014—2018 National Forest Resources Inventory Results *China Forest Resources Report*, China's forest coverage rate reached 22.96%.
- By the end of 2020, the national forest coverage rate reached 23.04%.
- The *National Major Ecosystem Protection and Restoration Major Project Master Plan (2021—2035)* put forward the target of 26% forest coverage in 2035.

- 根据 2014—2018 年的全国森林资源清查成果《中国森林资源报告》，中国森林覆盖率达 22.96%。
- 2020 年年底，全国森林覆盖率达 23.04%。
- 《全国重要生态系统保护和修复重大工程总体规划（2021—2035 年）》提出 2035 年森林覆盖率达到 26% 的目标。

9 National action for green and low-carbon
绿色低碳全民行动

Enhance national awareness of conservation, environmental protection, and ecology, advocate a simple, moderate, green, low-carbon, civilized and healthy lifestyle, and transform green concepts into conscious actions of all people.

增强全民节约意识、环保意识、生态意识，倡导简约适度、绿色低碳、文明健康的生活方式，把绿色理念转化为全体人民的自觉行动。

10 Orderly implement carbon peaking actions
有序实施碳达峰行动

- Determine carbon peaking goals scientifically and rationally
- Promote green and low-carbon development according to local conditions
- Coordinate every level of authorities to formulate local peaking plans
- 科学且合理地制定碳达峰目标
- 因地制宜推进绿色低碳发展
- 各级联动制定地方达峰方案

Source 资料来源：Circular of the State Council on an *action plan for peaking carbon emissions before 2030*
国务院关于印发《2030 年前碳达峰行动方案》的通知。

China's Climate Action
中国的气候行动

Green Policies in the "14th Five-Year" Plan/“十四五”规划中的绿色政策

The "14th Five-Year" Plan is the blueprint and action plan for China's economic and social development from 2021 to 2025. In the document, many green policies have been established to facilitate the achievement of carbon peak and carbon neutrality.

“十四五”规划是中国由2021年至2025年经济社会发展的蓝图和行动纲领，其中为达成碳达峰与碳中和目标设立了诸多绿色政策。

Implementing smart production and green production projects, fostering the high-end, smart and green development of the manufacturing industry
实施智能制造和绿色制造工程，
推动制造业高端化智能化绿色化 (3.8.3)

Launching national water conservation projects
实施国家节水行动 (11.39.1)

Establishing a unified system on the standards, certification and labelling of green products
建立统一的绿色产品标准、认证、标识体系 (11.39.3)

Formulating scientific plans for building urban green corridors, and building low-carbon cities
规划科学布局，以建设城市
绿色走廊及低碳城市 (8.29.2)

Actively advancing the development of green finance
致力发展绿色金融
(11.39.4)

Implementing tax policies conducive to energy conservation and the integrated utilisation of resources
实施有利于节能环保和资源综合利用的税收政策 (11.39.4)

Green Development of The Belt and Road Initiative
“一带一路”绿色发展

The Silk Road Economic Belt and the 21st-century Maritime Silk Road (The Belt and Road Initiative) is a global infrastructure development strategy adopted by the Chinese government in 2013 to invest in nearly 70 countries and international organizations.

丝绸之路经济带和21世纪海上丝绸之路（简称“一带一路”）是中国于2013年开始主导的跨国经济带，范围涵盖中国历史上的丝绸之路以及海上丝绸之路所经过的近70个国家和国际组织。

In April 2022, the *Guidance on Promoting the Green Development of the Belt and Road Initiative* was released, putting forward 15 specific tasks and key areas to promote green development, including green infrastructure interconnection, green energy, green transportation, green industry, green trade, green finance, green technology, green standards, and addressing climate change.

All parties were urged to fully implement the *United Nations Framework Convention on Climate Change* and its *Paris Agreement*.

2022年4月，《关于推进共建“一带一路”绿色发展的意见》发布，围绕推进绿色发展提出15项具体任务，包括绿色基础设施互联互通、绿色能源、绿色交通、绿色产业、绿色贸易、绿色金融、绿色科技、绿色标准、应对气候变化等重点领域，推动各方全面履行《联合国气候变化框架公约》及《巴黎协定》。

Source 资料来源：*Outline of the 14th Five-Year Plan (2021—2025) for National Economic and Social Development and Vision 2035 of the People's Republic of China.* 《中华人民共和国国民经济和社会发展第十四个五年规划和2035年远景目标纲要》。

China's Climate Action
中国的气候行动

Climate Action Case Studies/ 碳中和行动案例

Sustainability in the spotlight at Beijing Olympics 2022
2022 年北京绿色可持续冬奥

Vision :"Sustainability for the Future"

Objective : "Creating a New Example for Staging Events and Regional Sustainability"

Framework constructed from three aspects: Positive Environmental Impact, New Development for the Region, and Better Life for the People, including 12 actions, 37 tasks and 119 measures.

愿景：“可持续 · 向未来”

目标：“创造奥运和地区可持续发展新典范”

从环境正影响、区域新发展、生活更美好 3 个方面构建了工作框架，共包括 12 项行动、37 项任务和 119 条具体措施。

The Sustainability Policy for the Olympic and Paralympic Winter Games Beijing 2022 elaborates on priorities of ecosystem and biodiversity conservation and environmental management. On 30 September 2021, Beijing 2022 achieved and completed 98% of the sustainability objectives and tasks.

《北京 2022 年冬奥会和冬残奥会可持续性政策》落实了保护生态系统及生物多样性、开展环境管理等重点任务。截至 2021 年 9 月 30 日，北京冬奥会的可持续性承诺已完成 98%。

New Purposes for Existing Venues
赋予现有场地新功能

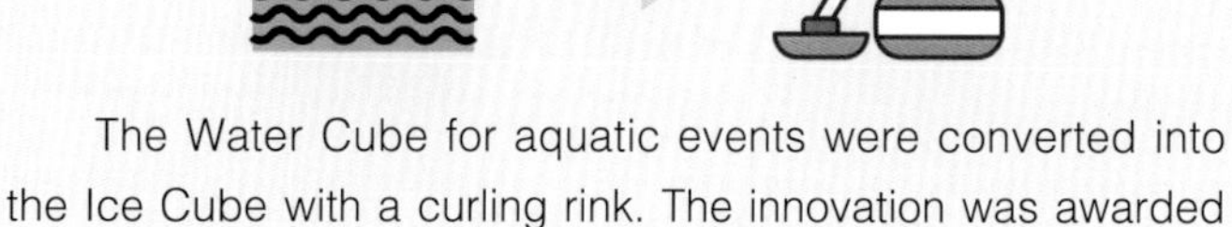

The Water Cube for aquatic events were converted into the Ice Cube with a curling rink. The innovation was awarded the 2019 IOC trophy "Sports and Sustainable Architecture".

国家游泳中心水上赛事比赛大厅转换成“冰立方”冰壶场地。该创新获得2019年国际奥委会颁发的“体育和可持续建筑”奖杯。

Wildlife Conservation in the Yanqing Competition Zones
延庆赛区野生动物保护

In the process of venue planning, design and constructions, a number of effective measures were taken to minimize the impacts on wild animals and their habitats in the core area of the Yanqing Zone. Match the measures with the correpsonding picture (Answers in next page).

延庆赛区核心区在场馆规划设计和建设过程中采取了多项措施，将对野生动物及其栖息地的影响降至最小。试着为以下措施找到相应的图片。（答案见下页）

1.

2.

3.

4.

A. More than 600 artificial nests were set up around the Zone
在赛区周边设置了 600 多个人工鸟巢

B. Adopted the DNA-based individual measurement method to monitor the population of animals (e.g. Ring-necked pheasant) around the Zone
采用 DNA 个体测定方法监测赛区周边动物（如环颈雉等）的种群分布

C. Protect nestling in artificial nests
保护人工鸟巢中的雏鸟

D. Infrared trigger cameras are set up to continuously monitor the activites of wild animals such as Gorals
设置红外触发相机，持续监测中华斑羚等野生动物的活动状况

Source 资料来源：Beijing Organising Committee for the 2022 Olympic and Paralympic Winter Games
北京 2022 年冬奥会和冬残奥会组织委员会。

China's Climate Action
中国的气候行动

Climate Action Case Studies/ 碳中和行动案例

Sustainability in the spotlight at Beijing Olympics 2022
2022 年北京绿色可持续冬奥

Minimizing Water Used for Snow Making

All snow sport venues have adopted a smart snowmaking system, with all equipment incorporated into a platform for unified management, enabling fast snow production, high snow-making efficiency and low water consumption. Compared with traditional methods, smart snow making can save up to 20% of water by realising optimal allocation and accurate delivery of water resources.

减少人工造雪用水量

北京冬奥会雪上场馆采用智能造雪系统，将所有造雪设备集中到一个平台进行统一管理。相较于传统造雪，智能造雪设备的优势在于造雪速度快、效率高且用水量少，能节约 20% 的水资源，可实现水资源优化配置和精准投放。

Promoting the Use of Low Carbon Energy

The Beijing 2022 Games relied on the newly-built Zhangbei renewable energy flexible DC power grid in Zhangjiakou City and it's cross-regional green power trading mechanism to meet the Games-time energy demands of all venues across the three competition zones. For the first time in Olympic history, 100% of the conventional electricity demand of all venues were supplied by renewable energy at Games-time.

促进低碳能源的使用

2022 年北京奥运会依托于张家口市新建的张北可再生能源柔性直流电网及其跨区域绿色电力交易机制，满足了三个赛区所有场馆在赛事期间的能源需求。这是奥运会历史上第一次所有场馆的常规电力需求在比赛期间 100% 由可再生能源提供。

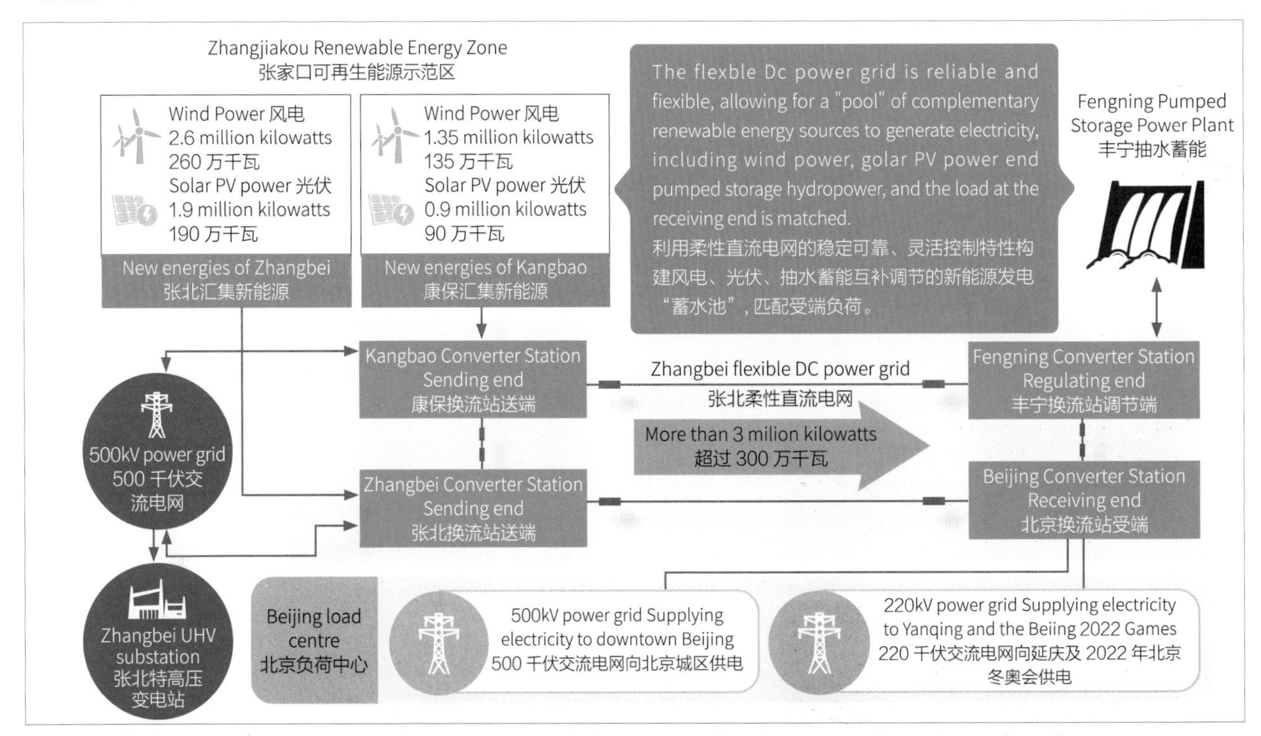

Illustrative diagram of renewable energy transmission for the 2022 Beijing Winter Olympics
2022 年北京冬奥会可再生能源传输示意图

Source 资料来源：Beijing Organising Committee for the 2022 Olympic and Paralympic Winter Games 北京 2022 年冬奥会和冬残奥会组织委员会。

Answer Key 答案（P.63）：1. A 3. B 2. D 4. C

China's Climate Action
中国的气候行动

Hong Kong's Climate Action/ 香港的气候行动

Over the last hundred years, the numbers of hot nights (days with a minimum temperature of 28°C or above) and very hot days (days with a maximum temperature of 33°C or above) in Hong Kong have increased while the number of cold days (days with a minimum temperature of 12°C or below) has decreased. In 2020, the number of very hot days and hot nights in Hong Kong were both the highest on record, and the summer was also the hottest on record.

过去 100 年间，香港热夜（日间最低气温 28℃或以上）的天数和酷热（日间最高气温 33℃或以上）的天数上升，寒冷（日间最低气温 12℃或以下）天数则下降。2020 年香港的酷热天数和热夜天数均是有记录以来最高的，同年夏季也是有记录以来最热的。

Impact of climate change to Hong Kong
气候变化对香港的影响

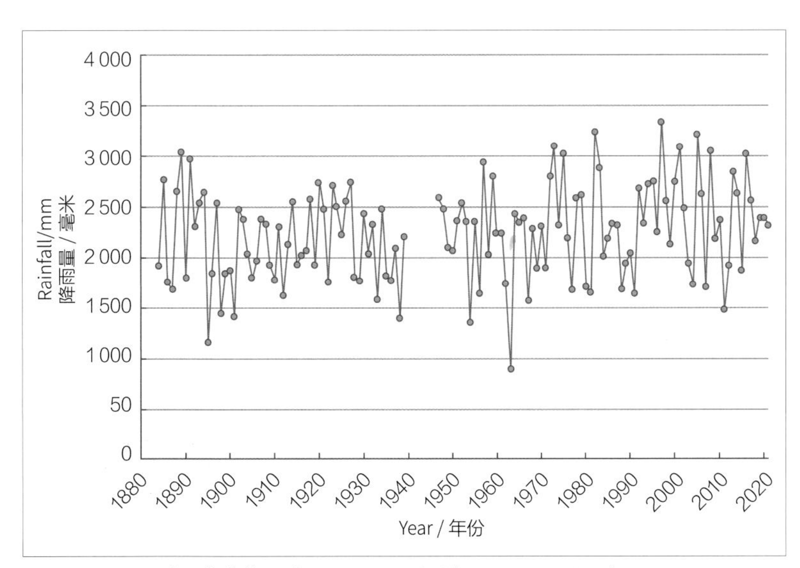

Annual Rainfall at the Hong Kong Observatory Headquarters
香港天文台总部的年降雨量

At the Hong Kong Observatory Headquarters, information for rainfall and heavy rain days (hourly rainfall greater than 30mm) have been available since January and March 1884 respectively, except for a break during World War II from 1940 to 1946.

Analysis of the annual rainfall showed that there was an average rise of 2.3 mm per year from 1884 to 2021.

For the annual number of heavy rain days, there was an average rise of 0.2 day per decade from 1884 to 2021.

除了在 1940—1946 年因第二次世界大战而中断外，香港天文台总部记录的降雨量及大雨天数（每小时降雨量超过 30 毫米）资料分别自 1884 年 1 月及 3 月开始。

分析显示，1884—2021 年，年降雨量的平均上升速度为每年 2.3 毫米。而 1884—2021 年，每年大雨天数的平均上升速度为每 10 年 0.2 天。

Source 资料来源：Hong Kong Observatory 香港天文台。

China's Climate Action
中国的气候行动

History of Sustainable Development in Hong Kong—Since 1999
香港可持续发展的历史——自 1999 年起

In his *Policy Address in 1999*, the Chief Executive set out his intention to build Hong Kong into a world-class city, and for the first time, sustainable development was put on the Government's agenda and was brought to the public's attention. Simply put, sustainable development for Hong Kong means:

香港特别行政区行政长官在《1999 年施政报告》中宣布，计划将香港建设成为世界级都会，首次将可持续发展纳入政府的工作日程并让公众知悉。简单来说，香港的可持续发展意味着：

Finding ways to increase prosperity and improve the quality of life while reducing overall pollution and waste

在追求经济繁荣、改善生活质量的同时，减少整体污染和浪费

Meeting our own needs and aspirations without doing damage to the prospects of future generations

在满足我们各种需要和期望的同时，不损害子孙后代的福祉

Reducing the environmental burden we put on our neighbours and helping to preserve common resources

减少对邻近区域造成环保负担，协力保护共同拥有的资源

From 2004 to 2022, Hong Kong conducted nine public engagements to collect public suggestions for Hong Kong to move towards carbon neutrality and sustainable development. Let's take a look at the topics that Hong Kong citizens are concerned about:

- Control of Single-use Plastics (Sep 2021—Apr 2022)
- Long-term Decarbonisation Strategy (Jun 2019—Nov 2020)
- Promotion of Sustainable Consumption of Biological Resources (Jul 2016—Nov 2017)
- Municipal Solid Waste Charging (Sep 2013—Apr 2015)
- Combating Climate Change: Energy Saving and Carbon Emission Reduction in Buildings (Aug 2011—Jun 2012)
- Building Design to Foster a Quality and Sustainable Built Environment (Jun 2009—Oct 2010)
- Clean Air - Clear Choices (Jun 2007—Oct 2008)
- Enhancing Population Potential for a Sustainable Future (Jun 2006—Dec 2007)
- Making Choices for Our Future (Jul 2004—May 2005)

"Building Design to Foster a Quality and Sustainable Built Environment" public engagement
“优化建筑设计，缔造可持续建筑环境” 公众参与

2004—2022 年，香港特别行政区政府进行了九次公众咨询，收集公众对于香港迈向碳中和和可持续发展的建议。让我们看看民众都关注些什么议题：

- 管制即弃塑料 (2021 年 9 月至 2022 年 4 月)
- 长远减碳策略 (2019 年 6 月至 2020 年 11 月)
- 推广可持续使用生物资源 (2016 年 7 月至 2017 年 11 月)
- 都市固体废物收费 (2013 年 9 月至 2015 年 4 月)
- 纾缓气候变化：从楼宇节能减排开始 (2011 年 8 月至 2012 年 6 月)
- 优化建筑设计，缔造可持续建筑环境 (2009 年 6 月至 2010 年 10 月)
- 未来空气 今日靠你 (2007 年 6 月至 2008 年 10 月)
- 为可持续发展提升人口潜能 (2006 年 6 月至 2007 年 12 月)
- 为我们的未来作出抉择 (2004 年 7 月至 2005 年 5 月)

Source 资料来源：Environment and Ecology Bureau of HKSAR 香港特别行政区环境及生态局。

"Long-term Decarbonisation Strategy" public engagement
“长远减碳策略” 公众参与

"Control of Single-use Plastics" public engagement
“管制即弃塑料” 公众参与

China's Climate Action
中国的气候行动

In October 2021, The Hong Kong Government announced *Hong Kong's Climate Action Plan 2050*, setting out the vision of "Zero-carbon Emissions · Livable City · Sustainable Development", and outlining the strategies and targets for combating climate change and achieving carbon neutrality.

In response to the *Paris Agreement*, the Government announced *Hong Kong's Climate Action Plan 2030+* in 2017, setting out the decarbonisation target of reducing Hong Kong's carbon intensity by 65% to 70% by 2030 using 2005 as the base, which is equivalent to a reduction in the total carbon emissions by 26% per cent to 36%. With the implementation of various mitigation measures, Hong Kong is moving steadily towards the 2030 decarbonisation target. The carbon intensity in 2019 was about 35% per cent lower than that in 2005. Preliminary estimation shows that the per capita carbon emissions reduced from the peak level of 6.2 tonnes in 2014 to about 4.55 tonnes in 2022.

The four major decarbonisation strategies in *Hong Kong's Climate Action Plan 2050* covers: Net zero electricity generation, Energy saving and green buildings, Green transport and Waste reduction.

香港特别行政区政府于2021年10月公布《香港气候行动蓝图2050》，以“零碳排放 · 绿色宜居 · 持续发展”为愿景，提出香港应对气候变化和实现碳中和的策略和目标。

为响应《巴黎协定》，香港特别行政区政府于2017年公布《香港气候行动蓝图2030+》，提出在2030年将本港碳强度由2005年的水平，降低65%～70%的减碳目标，相等于碳排放总量降低26%~36%。随着各项减缓措施相继落实，香港正逐步迈向2030年的减碳目标。2019年的碳强度较2005年下降约35%。据统计，2022年人均碳排放量已由2014年峰值的6.2吨降至4.55吨。

《香港气候行动蓝图2050》提出的四大减碳策略涵盖以下目标和措施：净零发电、节能绿建、绿色运输与全民减废。

香港气候行动蓝图 2050

2014 Peak 达峰

6.2 tonnes per capita CO_2e

人均 6.2 吨二氧化碳当量

2020 Progress 进展

4.5 tonnes per capita CO_2e

人均 4.5 吨二氧化碳当量

2035 −50% 减少 50%

2~3 tonnes per capita CO_2e

人均 2~3 吨二氧化碳当量

2050 carbon neutrality 达到碳中和

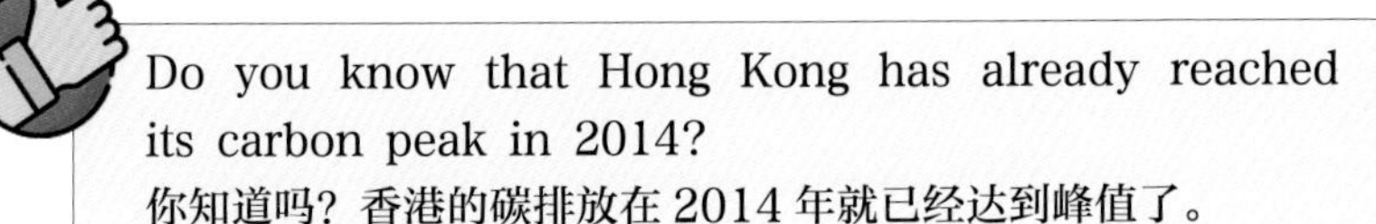

Do you know that Hong Kong has already reached its carbon peak in 2014?
你知道吗？香港的碳排放在2014年就已经达到峰值了。

Compare Hong Kong's net zero target and strategies with those of other countries and regions, for example, with China's 2060 net zero target, think about why different cities and countries set different goals considering social and economic structures.

将中国香港的碳中和目标和策略与其他地区相比，如与中国内地的2060碳中和目标进行对比，从社会、经济结构等方面进行思考，为何不同城市 / 国家设定不同的碳中和目标？

《香港清新空气蓝图 2035》

《香港资源循环蓝图 2035》

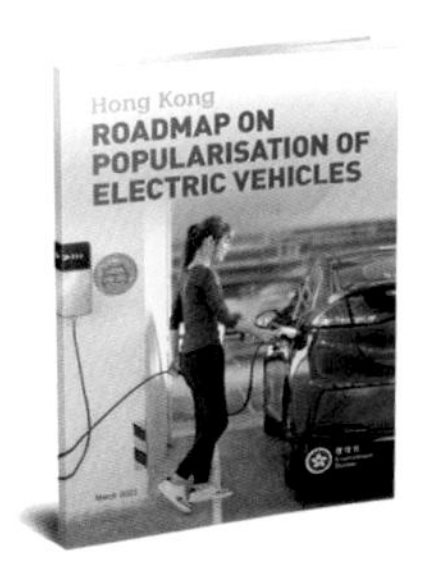

《香港电动车普及化路线图》

Source 资料来源：Environment and Ecology Bureau of HKSAR 香港特别行政区环境及生态局。

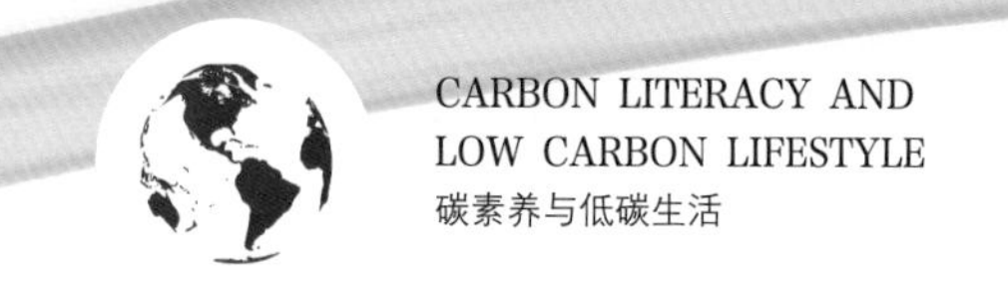

China's Climate Action
中国的气候行动

Electricity generation, transport and waste are the three major emission sources, together accounted for over 90% of the total emissions, and are therefore the three most critical areas of Hong Kong's decarbonisation work.

发电、运输和废弃物是占比最多的三大排放源，占总排放量的 90% 以上，也是香港碳中和工作最关键的 3 个领域。

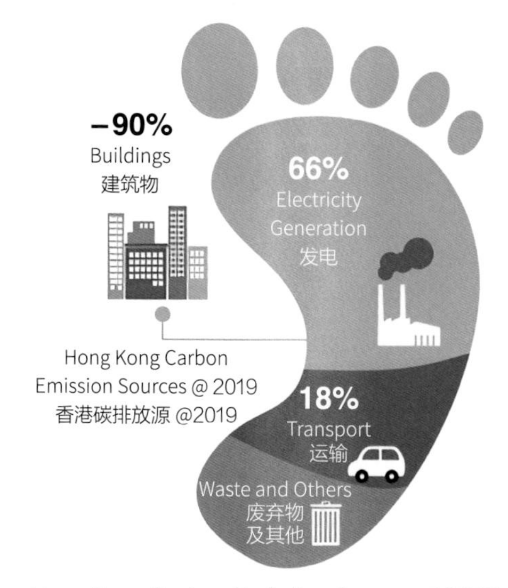

Hong Kong Carbon Emission Sources @2019
香港 2019 年碳排放源

Net Zero Electricity Generation 净零发电

Long-term target: Net zero carbon emissions in electricity generation before 2050

长远目标：2050 年前发电达到碳中和

Medium-term targets: Phasing out coal for electricity generation, Developing RE

中期目标：淘汰燃煤发电，发展可再生能源

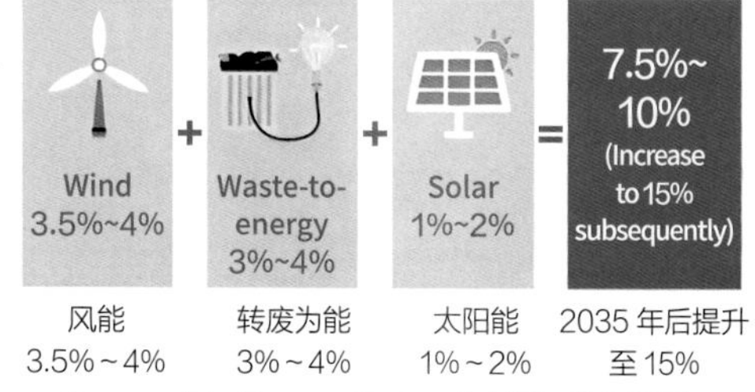

Renewable Energy Potential (Until 2035)
可再生能源的潜力（至 2035 年）

As technologies advanced rapidly, waste-to-energy technology in Hong Kong has been widely used. Let's take a look at a few cases:
随着科技进步，香港的转废为能技术已经被广泛运用。让我们一起了解几个案例：

The Food Waste/Sewage Sludge Anaerobic Co-digestion Trial Scheme at Tai Po Sewage Treatment Works (STW) (May 2019) receive food waste for food waste/sewage sludge anaerobic co-digestion. Apart from increasing the biogas yield and reducing the amount of digestate and carbon emissions from the Tai Po STW, the pilot scheme can also enhance Hong Kong's food waste treatment capacity and turn waste into electricity. Under this trial scheme, up to 50 tonnes of food waste can be treated per day, and the energy to be generated annually is estimated to be about 950 000 kW·h.

在香港大埔污水处理厂推行的“厨余、污泥共厌氧消化”试验计划于 2019 年 5 月开始接收厨余，将厨余与污泥进行共厌氧消化。此项计划除了可以增加生物气体产量及减低沼渣量、减少污水处理厂的碳排放之外，同时提升了香港的厨余处理能力，转废为能。试验计划每日可处理 50 吨厨余，预计每年可产生相等于约 95 万千瓦时电的能源。

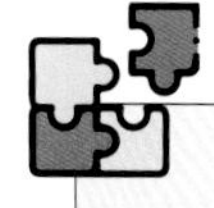

The T·Park Waste treatment facility was built in Nim Wan, Tuen Mun to support the "Sludge to Energy" Technology, visit its website and make a poster to explain the process of its technology.

在香港屯门稔湾的 T · Park 废物处理设施使用的是“污泥转化为能源”的技术。浏览其网站并制作海报来解释它所使用的技术和将污泥转变为能源的过程。

Source 资料来源：Environment and Ecology Bureau of HKSAR 香港特别行政区环境及生态局。

China's Climate Action
中国的气候行动

Energy Saving and Green Buildings
节能绿建

Long-term targets: Reduce the electricity consumption of commercial buildings by 30% to 40%, and that of residential buildings by 20% to 30% by 2050

长远目标：2050 年或之前，商业楼宇用电量减少 30%～40%，住宅楼宇用电量减少 20%～30%

Medium-term targets: Reduce the electricity consumption of commercial buildings by 15% to 20%, and that of residential buildings by 10% to 15% by 2035

中期目标：2035 年或之前，商业楼宇用电量减少 15%～20%，住宅楼宇用电量减少 10%～15%

Have you ever seen energy labels on the appliances in your home? Try to collect information about the energy efficiency of your home appliances, are you a energy saver?
你在家中的电器上见过能源标签吗？尝试收集家中电器的能源效益，成为节能小达人吧。

Appliance 电器	Energy Label Grade 能源标签级别
Refrigerator 冰箱	
Microwave 微波炉	
Television 电视机	
Air Conditioner 空调	
Washing Machine 洗衣机	

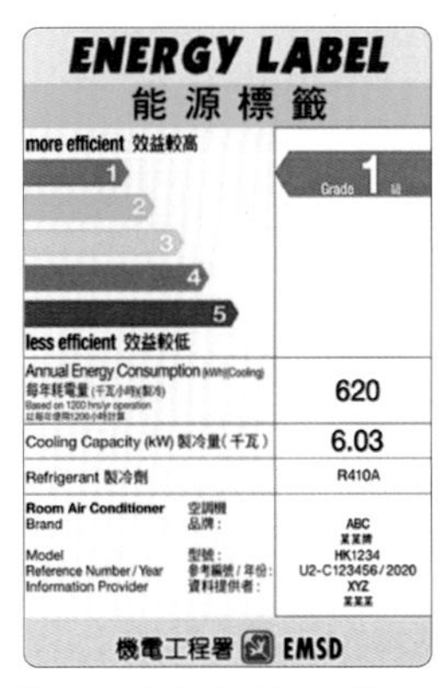

Energy label - Hong Kong
中国香港能源标签

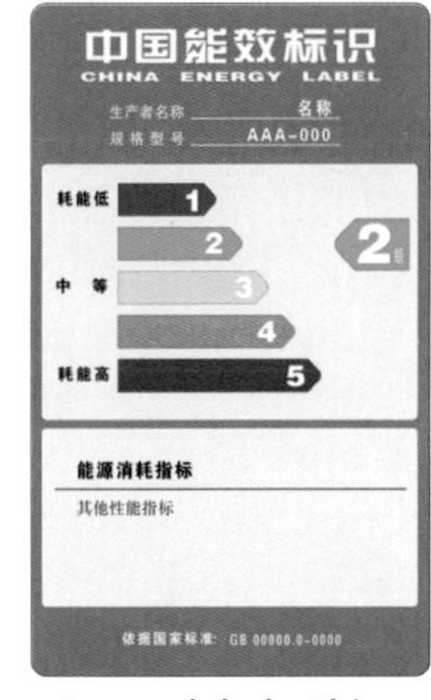

Energy label - China
中国能效标识

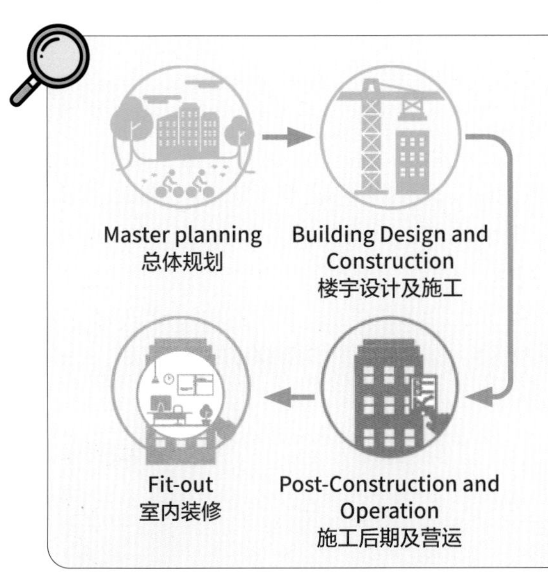

Hong Kong's very first green building benchmark was launched in 1996. BEAM Plus is the Hong Kong's leading initiative to offer independent assessments of building sustainability performance, offering a comprehensive set of performance criteria for a wide range of sustainability issues relating to the planning, design, construction, commissioning, fitting out, management, operation and maintenance of a building.

1996 年，中国香港首个绿色建筑认证系统“绿建环评”面世，该认证系统就建筑物在规划、设计、施工、调试、装修、管理、运作及维修中各范畴的可持续性，制定了一套全面的表现准则。评核结果受香港绿色建筑议会认可并获得认证。

Source 资料来源：Environment and Ecology Bureau of HKSAR 香港特别行政区环境及生态局。

China's Climate Action
中国的气候行动

3 Green Transport
绿色运输

Long-term target: Zero carbon emissions from vehicles and transport sector before 2050

长远目标：到 2025 年实现车辆和运输部门的零碳排放

Medium-term target: Set a concrete timetable for adopting new energy public transport

中期目标：确立具体使用新能源公共交通工具时间表

Electric Vehicles in Hong Kong
香港电动车

In the first 5 months of 2021, the proportion of electric private cars in newly registered private cars has further increased from 12.4% in 2020 to 18.4%, representing that more than 1 out of every 6 new private cars is electric.

在 2021 年的前 5 个月，电动私家车登记占新登记私家车的比率由 2020 年的 12.4% 上升至 18.4%，即每 6 辆新私家车中至少有一辆为电动车。

In October 2020, Hong Kong Launched the $2 billion "EV-Charging at Home Subsidy Scheme" to subsidise installation of EV charging infrastructure in car parks of existing private residential buildings. After one year of implementation of the scheme, over 440 applications were received, covering nearly 100,000 parking spaces, ensuring sufficient and convinient charging infrastructure for EV owners.

香港在 2020 年 10 月推出总额 20 亿港币的“EV 屋苑充电易资助计划”，资助现有私人住宅楼宇停车场安装电动车充电基础设施。计划实施一年来，已收到超过 440 份申请，涉及近 100 000 个车位，以确保电动车充电设施供应充足、方便。

4 Waste Reduction
全民减废

Long-term Target: Carbon Neutrality in Waste Management

长远目标：废物处理达到碳中和

Medium-term Target: Enhance Waste Reduction and Recycling

中期目标：加强减废和回收

Through implementing Municipal Solid Waste (MSW) charging and other waste reduction and recycling initiatives, and encouraging the whole community to work together, we aim to progressively reduce the per capita MSW disposal by 40% to 45% and raise the recovery rate to about 55%.

香港正努力通过推行都市固体废物征费及其他减废回收措施，鼓励全民参与，目标是把都市固体废物的人均弃置量逐步减少 40%～45%，同时把回收率提升至约 55%。

Not all paper and plastic waste are suitable for recycling! Do some research to learn about which ones shouldn't be brought to the recycling station.

并不是所有废纸和塑料废物都适合回收！通过网上资料搜索，了解哪些不适合带到回收站。

Source 资料来源：Environment and Ecology Bureau of HKSAR 香港特别行政区环境及生态局。

Individual Strategies to Combat Climate Change
个人应对气候变化的策略

Everyone could take climate actions. Starting from ourselves, we could then call on the people around us to take actions together to create a sustainable future.

每个人都可以为应对气候变化做出行动。我们应从自身做起，然后呼吁身边的人与我们一同创造可持续的未来。

1. Save energy at home 居家节能

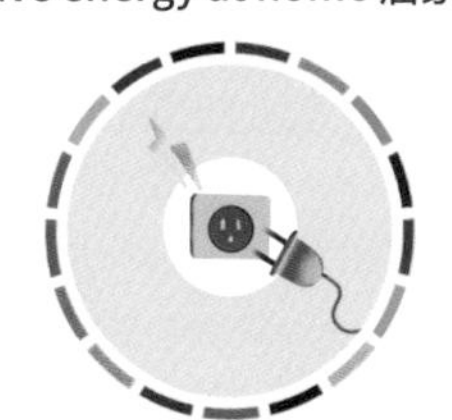

Use less energy by reducing your heating and cooling use, switching to LED light bulbs and energy-efficient electric appliances, washing your laundry with cold water, or hanging things to dry instead of using a dryer. Improving your home's energy efficiency, through better insulation for instance, or replacing your oil or gas furnace with an electric heat pump can reduce your carbon footprint by up to 900 kilograms of CO_2e per year.

通过减少使用供暖设备或空调设备、改用 LED 灯泡和节能电器、用冷水洗衣服、衣物以晾晒代替使用烘干机等方式，人们可以减少能源消耗。通过提升房屋的隔热性、使用电热泵代替油炉和燃气炉等方式来提高居家能源利用效率，每年可减少 900 千克二氧化碳当量。

2. Walk, bike or take public transport 步行、骑行或者乘坐公共交通

Walking or riding a bike instead of driving will reduce greenhouse gas emissions and help your health. For longer distances, consider taking a train or bus. Living car-free can reduce your carbon footprint by up to 2 tons of CO_2e per year compared to a lifestyle using a car.

以步行或骑行代替驾车出行不仅可以减少温室气体排放，还有助于保持身体健康。如果距离较远，可以考虑乘坐火车或公共汽车。与驾车出行的生活方式相比，这样每年可以减排多达 2 吨二氧化碳当量。

3. Reduce, reuse, repair and recycle 减少、再利用、修理、回收

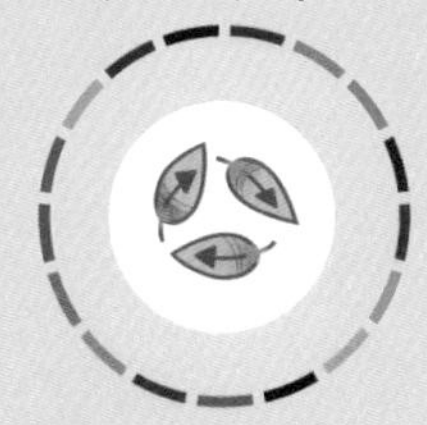

Reduce unnecessary shopping, shop second-hand, and repair what you can. Every kilogram of textiles produced generates about 17 kilograms of CO_2e, so buying fewer new clothes and other consumer goods can also reduce your carbon footprint.

我们可以通过选择二手商品，对可修理的物品进行修理并积极进行废物回收来减少不必要的购物。因为每生产 1 千克的纺织品就会产生大约 17 千克二氧化碳当量，所以我们应该尽量少买新衣服和其他消费品，以减少碳足迹和浪费。

4. Consider your travel 选择低碳的旅行方式

Taking one less long-haul return flight can reduce your carbon footprint by up to almost 2 tons of CO_2e. When you can, meet virtually, take a train, or skip that long-distance trip altogether.

少坐一次长途回程航班可减少二氧化碳当量约 2 吨。如果情况允许，尽量乘坐火车，或利用网络会议以减少长途旅行。

Source 资料来源：UN 联合国。

Individual Strategies to Combat Climate Change
个人应对气候变化的策略

5. Eat more vegetables 多吃蔬菜

Eating more vegetables, fruits, whole grains, legumes, nuts, and seeds, and less meat and dairy, can significantly lower your environmental impact. Producing plant-based foods generally results in fewer greenhouse gas emissions and requires less energy, land, and water. Shifting from a mixed to a vegetarian diet can reduce your carbon footprint by up to 500 kilograms of CO_2 e per year.

多吃蔬菜、水果、全谷物、豆类、坚果和籽类，少食用肉类和乳制品，这样可以大大降低个人对环境的影响。生产蔬果所需的能源、土地、水资源均较少，排放的温室气体也通常更少。例如，若将荤素混合的饮食结构转为素食，每年可减少 500 千克二氧化碳当量的碳足迹。

6. Clean up your environment 净化环境

Every year, people throw out 2 billion tons of trash. About a third causes environment harms, from choking water supplies to poisoning soil. Dispose waste and garbages properly and educate others to do the same, or participate in local clean-ups of parks, rivers, beaches and beyond.

每年，人们扔掉 20 亿吨垃圾，其中大约 1/3 的垃圾对环境造成了如堵塞水供应渠道、破坏土壤等危害。因此请妥善处理废物和垃圾，并倡导身边的人也这样做。我们也可以参与所在地的清理公园、河流、海滩等活动。

7. Choose eco-friendly products 选择生态友好型产品

Everything we spend money on affects the planet. You have the power to choose which goods and services you support. To reduce your environmental impact, choose products from companies who use resources responsibly and are committed to cutting their gas emissions and waste.

虽然我们花钱购买的所有产品都会对地球的环境产生影响，但是我们有权选择我们支持的那些产品和服务。要选择那些负责任地使用资源并致力于减少温室气体排放和废物的公司生产的产品，这可以有效减少人们对环境的影响。

8. Speak up 呼吁

Talk to your neighbors, colleagues, friends, and family. Let business owners know your demand for low, carbon products. Climate action is a task for all of us, no one can do it all alone-but we can do it together.

呼吁你的邻居、同事、朋友和家人采取气候行动。让商家了解到人们对低碳产品的诉求。气候行动，人人有责，没有人能置身事外。单凭一己之力无法达成的应对气候变化的目标，需要人们众志成城，携手共进。

Source 资料来源：UN 联合国。

Individual Strategies to Combat Climate Change
个人应对气候变化的策略

My Climate Action Plan/ 我的气候行动计划

Overall Target 总目标

Monthly Targets 月度目标

Annual Targets 年度目标

Specific Actions 具体行动

- [ ]
- [ ]
- [ ]
- [ ]
- [ ]
- [ ]
- [ ]
- [ ]
- [ ]
- [ ]

Evaluation Method 评估方法

APPENDIX
附件

Greenhouse Gas Accounting
温室气体排放计算

Major Steps for GHG Accounting/ 碳审计的主要步骤

Step 1: Setting organizational and operational boundaries

Step 2: Identify base year to track and analyse emission trends over time

Step 3: Identify emission sources

Step 4: Select calculation approach

Step 5: Collect data and choose emission factors

Step 6: Apply calculation tools and roll-up data

Step 7: Accounting, Reporting and Verification

Step 8: Setting greenhouse gases emissions targets

Source 资料来源： World Resources Institute 世界资源研究所。

The following carbon emission calculation methods took reference from *The GHG Protocol Corporate Accounting and Reporting Standard (Revised Edition)* , but are also applicable to carbon footprint calculations for other types of organizations and individuals.

以下参考《温室气体核算体系企业核算与报告标准（修订版）》的碳排放计算方法也适用于其他类型的机构以及个人碳足迹计算。

第 1 步：设定组织 / 运营边界

第 2 步：选择基准年以跟踪和比较排放量的长期趋势

第 3 步：识别温室气体排放源

第 4 步：选择计算方法

第 5 步：收集数据并选择排放系数

第 6 步：运用计算工具，汇总数据

第 7 步：核算、报告、核查

第 8 步：设定温室气体排放目标

Globally recognised GHG Accounting Reference Documents
国际认可的碳审计参考文件

《温室气体核算体系
企业核算与报告标准（修订版）》
经济科学出版社，2022

The GHG Protocol Corporate Accounting and Reporting Standard (Revised Edition), 2004

The Product Life Cycle Accounting and Reporting Standard, 2011
《产品生命周期碳排放计算报告标准，2011》

ISO 14064–1: 2018, Greenhouse gases — Part 1: Specification with guidance at the organization level for quantification and reporting of greenhouse gas emissions and removals

ISO 14064–1: 2018，温室气体——第一部分：在组织层面对温室气体排放和减除进行量化和报告的规范和指引

Greenhouse Gas Accounting
温室气体排放计算

Key Principles in GHG Accounting/ 碳审计的主要原则

Relevance 相关性

Relevance means containing the information that users—both internal and external to the organization—need for their decision making. An important aspect of relevance is the selection of an appropriate inventory boundary that reflects the substance and economic reality of the organization's relationships, not merely its legal form.

一个企业的温室气体报告具备相关性，是指它包含企业内部和外部的用户做决策所需的信息。相关性的一个重要方面是选择适当的排放清单边界，这个边界应当反映该企业业务关系的本质和经济状况，而不只是它的法律形式。

Accuracy 准确性

GHG measurements, estimates, or calculations should be systemically neither over nor under the actual emissions value, as far as can be judged. The quantification process should be conducted in a manner that minimizes uncertainty.

在可知的范围内，应尽量使温室气体的测量、估算和计算不系统性地高于或低于实际排放值，并在可行的范围内最大限度地减少不确定性。量化的计算方法应最大限度地降低不确定性。

Completeness 完整性

All relevant emissions sources within the chosen inventory boundary need to be accounted for to compile a comprehensive and meaningful inventory.

为了编制一份全面且有意义的排放清单，选定的排放清单边界内的所有相关排放源都应予以核算。

Consistency 一致性

Users of GHG information will want to track and compare GHG emissions information over time in order to identify trends and to assess the performance of the reporting organization. The consistent application of accounting approaches, inventory boundary, and calculation methodologies is essential to producing comparable GHG emissions data over time.

使用者需要不断跟踪和比较温室气体排放信息，以便识别报告企业的发展情况，评价其绩效。采用一致的核算方法、排放清单边界和计算方法学，对获得长期可比较的温室气体排放数据至关重要。

Transparency 透明度

Transparency relates to the degree to which information on the processes, procedures, assumptions, and limitations of the GHG inventory are disclosed in a clear, factual, neutral, and understandable manner based on clear documentation and archives (i.e., an audit trail). Information needs to be recorded, compiled, and analyzed in a way that enables internal reviewers and external verifiers to attest to its credibility. Specific exclusions or inclusions need to be clearly identified and justified, assumptions disclosed, and appropriate references provided for the methodologies applied and the data sources used.

透明度与信息披露程度有关，指有关温室气体排放清单的工艺、程序、假设条件和局限性的信息，应根据清楚的记录和档案（即审计线索），以清晰、真实、中立和易懂的方式予以披露。信息记录、整理和分析的方法应使内部审查人员和外部核查人员可以证实其可信度。特殊的排除或计入事项要明确指出并说明理由，要披露假设条件，对所用的方法学和引用的数据要提供相应的参考文献。

Source 资料来源：World Resources Institute 世界资源研究所。

Greenhouse Gas Accounting
温室气体排放计算

Step 1: Setting Organizational and Operational Boundaries
第 1 步：设定组织 / 运营边界

Organizational Boundary: Two distinct approaches can be used to consolidate GHG emissions: the equity share and the control approaches.

组织边界：有两种不同的温室气体排放量合并方法可供选择：股权比例法和控制权法。

1. Equity Share Approach: The company accounts for GHG emissions from operations according to its share of equity in the operation.

股权比例法：企业根据其在业务中的股权比例核算温室气体排放量。

2. Control Approach: a company accounts for 100% of the GHG emissions from operations over which it has control. It does not account for GHG emissions from operations in which it owns an interest but has no control.

控制权法：公司对其控制的业务范围内的全部温室气体排放量进行核算，对其享有权益但不持有控制权的业务产生的温室气体排放量不进行核算。

Financial Control: Emissions from joint ventures where partners have joint financial control are accounted for based on the equity share approach.

财务控制权：对享有共同财务控制权的合资企业的排放量应按股权比例核算。

Operational Control: a company accounts for 100% of emissions from operations over which it or one of its subsidiaries has operational control.

运营控制权：公司对其自身或其子公司持有运营控制权的业务产生的 100% 的排放量进行核算。

Operational Boundary is usually classified into three separate scopes as below:

- Scope 1: Direct emissions and removals
- Scope 2: Electricity indirect GHG emissions
- Scope 3: Other indirect emissions

运营边界通常分为以下 3 个不同的范围：

- 范围 1：直接温室气体排放
- 范围 2：电力产生的间接温室气体排放
- 范围 3：其他间接温室气体排放

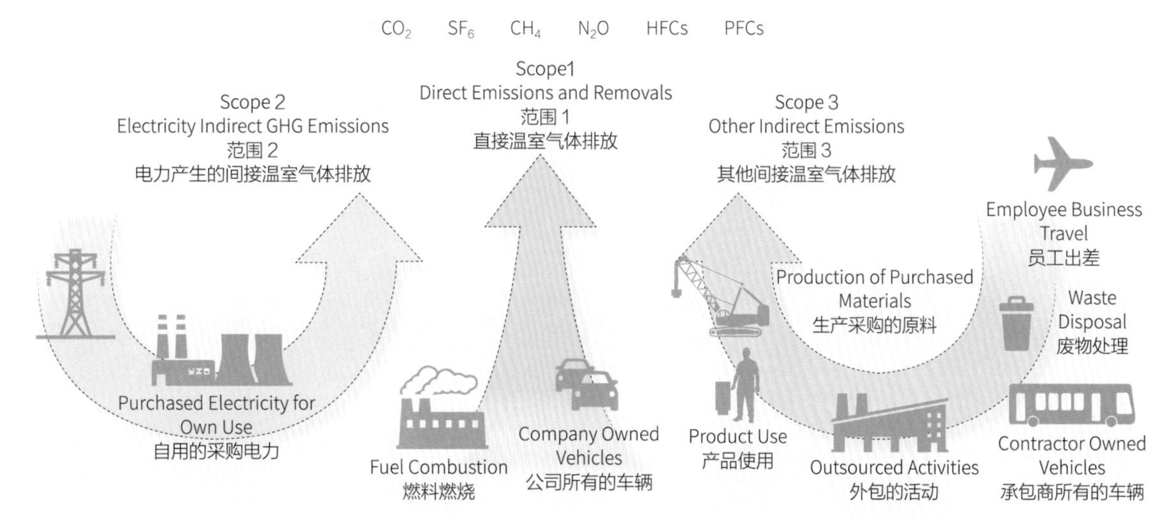

Overview of Emission Scopes on the Value Chain/ 价值链上的范围与排放概览

Source 资料来源：World Resources Institute 世界资源研究所。

Greenhouse Gas Accounting
温室气体排放计算

Step 2: Identify Base Year to Track and Analyse Emission Trends Over Time
第 2 步：选择基准年以跟踪和比较排放量的长期趋势

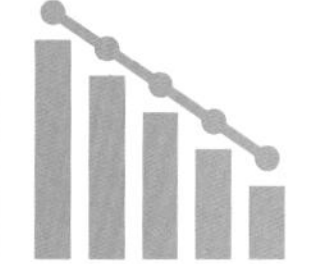

Selecting a Base Year

- A historic year or group of years against which an company's emissions can be tracked over time.
- Often the first year that an entity accounts for its GHG emissions.
- Must have verifiable emissions data available.
- Establishing a base year helps you with:
  Public reporting
  Establishing GHG targets
  Managing risks and opportunities
  Addressing the needs of stakeholders of the company

选择基准年

- 公司需要设定一个年份或几年间的平均排放量作为跟踪减排量的参照值。
- 通常是开始核算其温室气体排放的第一年。
- 基准年须有可供核查的排放数据。
- 建立基准年有助于：
  进行公开报告
  设定温室气体排放目标
  管理风险与机会
  满足投资者和其他利益相关方的需要

Step 3: Identify Emission Sources
第 3 步：识别温室气体排放源

Scope1: Direct GHG Emission
范围 1：直接温室气体排放

Companies report GHG emissions from sources they own or control as scope 1. Direct GHG emissions are principally the result of the following types of activities undertaken by the company.

- Generation of electricity, heat, or steam: These emissions result from combustion of fuels in stationary sources, e.g., boilers, furnaces, turbines.
- Physical or chemical processing: Most of these emissions result from manufacture or processing of chemicals and materials, e.g., cement, aluminum, adipic acid, ammonia manufacture, and waste processing.
- Transportation of materials, products, waste, and employees: These emissions result from the combustion of fuels in company owned/controlled mobile combustion sources (e.g., trucks, trains, ships, airplanes, buses, and cars).
- Fugitive emissions: These emissions result from intentional or unintentional releases, e.g., equipment leaks from joints, seals, packing, and gaskets; methane emissions from coal mines and venting; hydrofluorocarbon (HFC) emissions during the use of refrigeration and air conditioning equipment; and methane leakages from gas transport.

各公司在范围 1 中，报告其拥有或控制的排放源的温室气体排放情况。直接温室气体排放主要是公司从事下列活动产生的。

- 生产电力、热力或蒸汽：这些排放来自如锅炉、熔炉和涡轮机等固定排放源的燃料燃烧。
- 物理或化学工艺：这些排放主要来自化学品和原料的生产或加工，例如生产水泥、铝、己二酸、氨以及废弃物处理。
- 运输原料、产品、废弃物和员工：这些排放来自公司拥有 / 控制的运输工具（如卡车、火车、轮船、飞机、公共汽车和轿车）的燃烧排放源。
- 无组织排放：这些排放来自有意或无意的泄漏，例如，设备的接缝、密封件、包装和垫圈的泄漏，煤矿矿井和通风装置排放的甲烷，使用冷藏和空调设备过程中产生的氢氟碳化物（HFC）排放，以及天然气运输过程中的甲烷泄漏。

Source 资料来源：World Resources Institute 世界资源研究所。

Greenhouse Gas Accounting
温室气体排放计算

Most stationary combustion devices can be classified into the following categories.
大部分固定源的燃烧装置可归为以下其中一个类别。

Boilers
锅炉

Burners
火炉

Turbines
涡轮机

Internal Combustion Engines (e.g. Emergency Electricity Generator)
内燃机（例如紧急发电机）

Heaters
加热器

Furnace
熔炉

Oven
烤炉

Dryers
风干机

Any other equipment or machinery that combusts carbon bearing fuels or waste streams
其他燃烧含碳燃料或废物的设备或机械

Scope 2: Electricity Indirect GHG Emissions
范围 2：电力产生的间接温室气体排放

- Companys report the emissions from the generation of purchased electricity that is consumed in its owned or controlled equipment or operations as scope 2.
- Scope 2 emissions are a special category of indirect emissions.
- For many companys, purchased electricity represents one of the largest sources of GHG emissions and the most significant opportunity to reduce these emissions.
- Accounting for scope 2 emissions allows companies to assess the risks and opportunities associated with changing electricity and GHG emissions costs.
- 各公司在范围 2 中报告由其拥有或控制的设备或运营消耗的外购电力所产生的排放。
- 范围 2 的排放是一类特殊的间接排放。
- 对许多公司而言，外购电力是其最大的温室气体排放源之一， 也是减少其排放的最主要机会。
- 各公司通过核算范围 2 的排放，可以评估改变用电方式和温室气体排放成本的相关风险与机会。

Scope 2 usually includes purchased electricity and purchased heat, its emission factor should refer to the latest value published by government authorities.
外购电力及外购热力的排放因子应参考国家最新发布值。

Item 名称	Unit 单位	Default Value 缺省值
Electricity Emission Factor 电力排放因子	$tCO_2/MW \cdot h$	Use latest value published by government authorities 采用国家最新发布值
Heat Emission Factor 热力排放因子	tCO_2/GJ	0.11

Emission Factor Default Value for Purchased Electricity and Heat
外购电力、热力排放因子缺省值

Source 资料来源：World Resources Institute 世界资源研究所。

Greenhouse Gas Accounting
温室气体排放计算

Scope 3: Other Indirect Emissions (Optional for Reporting Purposes)
范围 3: 其他间接温室气体排放（可选择性报告）

- Scope 3 is optional, but it provides an opportunity to be innovative in GHG management. Companys may want to focus on accounting for and reporting those activities that are relevant to their goals, and for which they have reliable information.
- Since companys have discretion over which categories they choose to report, scope 3 may not lend itself well to comparisons across companys.
- 范围 3 是选择性的，但是它为创新性的温室气体管理提供了机会。各企业可能会重点关注核算和报告那些与其业务和目标相关的活动，以及那些有可靠信息的活动。
- 由于公司有权决定选择哪类信息进行报告，因此可能不能用范围 3 来对不同公司进行比较。

Examples of Scope 3 Emission Sources
范围 3 排放例子

- Extraction and production of purchased materials and fuels
- Transportation of purchased materials or goods
- Transportation of purchased fuels
- Employee business travel
- Transportation of waste
- Leased assets, franchises, and outsourced activities—emissions from such contractual arrangements are only classified as scope 3 if the selected consolidation approach (equity or control) does not apply to them
- Use of sold products and services
- Waste disposal

- 外购原料与燃料的开采和生产
- 采购材料或货物的运输
- 运输外购的燃料
- 职员差旅
- 运输废弃物
- 租赁资产、特许和外包活动——如果选定的合并方法（股权法或控制权法）不适用于这些合同活动，则它们产生的排放量只能归入范围 3
- 使用售出的产品和服务
- 废弃物处理

Imagine you are conducting Greenhouse Gas Accounting for a school, please classify the following sources into scopes 1, 2, and 3. (See next page for answers)
假设你正在为一所学校进行碳排放计算，请将以下排放源分为范围 1、范围 2 和范围 3。（答案见下页）

- Fuel consumption of school-owned school buses 学校拥有的校车所消耗的燃油
- Purchased electricity consumed by classroom lighting 学校教室照明消耗的购买的电力
- School teachers and staff commuting 学校教师与员工通勤交通
- Bunsen burner in laboratory 实验室本生灯的燃烧
- Air conditioner refrigerant used in classrooms 学校教室空调制冷剂的使用
- Printing paper usage 打印纸的使用
- Wastewater from school canteen 学校食堂产生的废水

Source 资料来源：World Resources Institute 世界资源研究所。

Greenhouse Gas Accounting
温室气体排放计算

Use of Carbon Calculation Tool & Apply Emission Factors/ 使用碳排放计算工具和排放系数计算

Activity Data (AD)
活动数据

Example

Annual energy consumption (kW·h/a)

Annual paper disposed to landfills (kg/a)

Annual water consumption (m^3/a)

例子

每年能源消耗（千瓦时 / 年）

每年填埋处理的纸张（千克 / 年）

每年用水量（米 ^{3}/ 年）

Emission Factors (EF)
排放因子

Emission factors are calculated ratios relating GHG emissions to a proxy measure of activity at an emissions source.

排放因子是经过计算得出的排放源活动水平与温室气体排放量之间的比率。

Example

2.614kgCO_2/kg diesel

0.023,9g methane/kg diesel

例子

2.614 千克二氧化碳 / 千克柴油

0.023 9 克甲烷 / 千克柴油

2

Note: Emission factors vary by region and industry, please refer to the guidelines issued by local authorities.

注：排放因子因地区及行业而异，请参考由本地权威机关发出的指引。

Global Warming Potential (GWP)
全球变暖潜能

A measure of how much energy the emissions of 1 ton of a gas will absorb over a given period of time (usually 100 years), relative to the emissions of 1 ton of carbon dioxide (CO_2).

衡量一种气体 1 吨的排放量在给定时间段（通常为 100 年）内相对于 1 吨二氧化碳 (CO_2) 能吸收多少能量。

*Please refer to P.10 for the GWP value

*GWP 数值表格请参考第 10 页

3

Carbon Dioxide Equivalent (CO_2e)
二氧化碳当量

GHG emissions are often measured in CO_2e. To convert emissions of a gas into CO_2 equivalent, its emissions are multiplied by the gas's Global Warming Potential (GWP).

Unit: tonne CO_2e/a

温室气体排放量通常以二氧化碳当量来衡量。要将气体的排放量转换为二氧化碳当量，其排放量乘以气体的全球变暖潜能值 (GWP)。

单位：吨二氧化碳当量 / 年

Why do GWPs change over time?

GWP values are updated occasionally. This change can be due to updated scientific estimates of the energy absorption or lifetime of the gases or to changing atmospheric concentrations of GHGs that result in a change in the energy absorption of 1 additional ton of a gas relative to another.

为什么 GWP 会随着时间变化而变化?

GWP 值会经常更新。产生这种变化可能是由于用于估算气体吸收能量的能力或其寿命的技术不断进步，或是由于大气中温室气体浓度的变化导致一种气体相对于另一种气体的能量吸收能力发生变化。

Source 资料来源：World Resources Institute 世界资源研究所。

Estimate the annual GHG emissions in the unit of kgCO_2e by a vehicle consuming an average of 167.5 L diesel oil per month. (See next page for answers)

估算一辆平均每月消耗 167.5 L 柴油的汽车的年度温室气体排放量，单位为 kgCO_2e。（答案见下页）

GHG 温室气体	EF 排放因子	Unit 单位
CO_2	2.614	kg/L
CH_4	0.072	g/L
N_2O	0.110	g/L

Diesel Oil emission factor (for reference only) 柴油排放因子（仅供参考）

Answer Key 答案（P.80）： 1、2、3、1、1、3、3

Greenhouse Gas Accounting
温室气体排放计算

Greenhouse Gas Management/ 温室气体管理

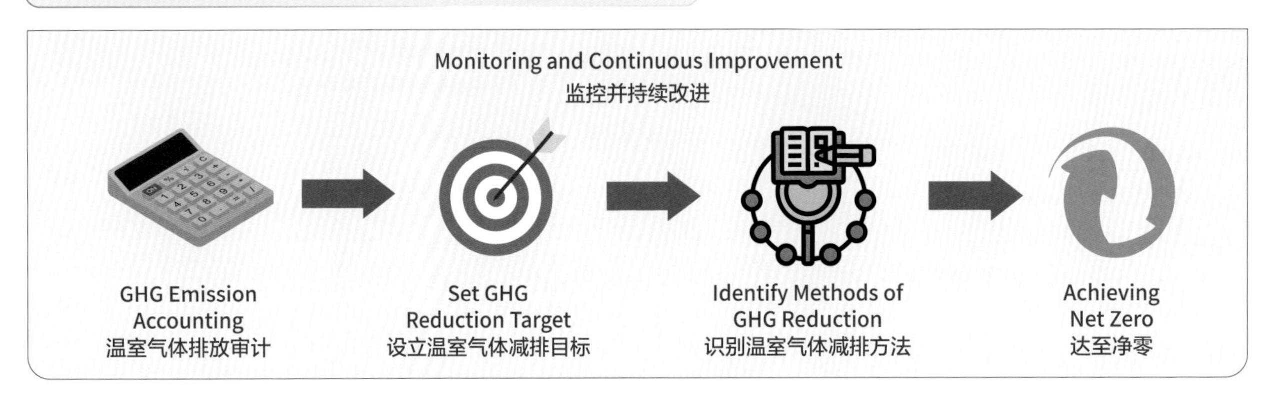

The ISO (the International company for Standardisation) specifies a "Plan-Do-Check-Act" (PDCA) management framework to implement carbon and energy management practices.

国际标准化组织 (ISO) 详细说明了如何使用“策划—实施—检查—改进”PDCA 的管理框架，把碳排放及能源管理带到日常实践中。

PLAN 策划

STEP 1: Establishing a Carbon Mangement Policy

- Demonstrate the companyal commitment to carbon management
- Set objectives and targets for improvement against the baseline

STEP 2: Establishing a Carbon Reduction Plan

- A plan to achieve the targets, which may include better management practices, minor hardware retrofitting and engineering improvement works

步骤 1：拟定碳管理政策

- 展示机构对碳管理的承诺
- 设定目标及指标，并根据基准做出改善

步骤 2：设立减碳计划

- 设立计划以达至减碳目标，当中可能包括更好的管理实践、少量硬件加装及工程改善工作

DO 实施

STEP 3: Implementing the Carbon Reduction Plan

步骤 3：实践减碳计划

CHECK 检查

STEP 4: Conducting Regular Carbon Audit

- Systematic procedures for monitoring of carbon emissions and the effectiveness of reduction measures
- Make adjustment when the company is not progressing well towards the objectives

步骤 4：定期进行碳审计

- 进行有系统的程序以监测碳排放及减碳措施的效能
- 如机构未能达到减碳目标，计划便需做出调整

ACT 改进

STEP 5: Maintaining the Carbon Reduction Plan

- Audit findings should be reviewed by management to ensure the measures are adequate and effective for continual improvement
- Communicate reduction success with stakeholders

步骤 5：维持实践减碳计划

- 管理层应就审计的调查结果进行研究分析，确保减碳措施足够及有效，已进行持续性的改善工作
- 多与持份者沟通，让他们了解你减碳工作的成效

Source 资料来源：ISO 国际标准化组织。

Answer Key 答案（P.81）：
167.5 × 12 = 2 010 L/a
CO_2: 2 010 × 2.614 × 1 = 5 254 kg
CH_4: 2 010 × 0.072 × 28/1000 = 3.329 kgCO_2e
N_2O: 2 010–0.110 × 265/1 000 = 65.44 kg CO_2e
total = 5 322 kgCO_2e

Greenhouse Gas Accounting
温室气体排放计算

Reporting Tool: Summary Table (Simplified Version for demonstration purposes)
报告工具：汇总表格（简易版，做示范用途）

Sources 排放源	CO_2	CH_4	N_2O	HFCs	PFCs	Subtotal 小计
Scope 1 范围 1						
Stationary Combustion Sources 固定排放源				N/A	N/A	
Mobile Combustion Sources 移动排放源				N/A	N/A	
Fugitive Emissions 逃逸性排放	N/A	N/A	N/A		N/A	
Other Direct Emissions 其他直接排放						
Scope 1 GHG Emission Total 范围 1 排放总计						
Planting of Additional Trees 植树		N/A	N/A	N/A	N/A	
Others 其他						
Scope 1 GHG Removal Total 范围 1 减除总计						

Greenhouse Gas Accounting
温室气体排放计算

Reporting Tool: Summary Table (Simplified Version for demonstration purposes)
报告工具：汇总表格（简易版，做示范用途）

Sources 排放源	Subtotal 小计
Scope 2 范围 2	
Electricity Purchased 购买电力	
Gas Purchased (for heat) 购买煤气（热力）	
Total Scope 2 GHG Emissions 范围 2 碳排放总计	

Sources 排放源	CO_2	CH_4	N_2O	HFCs	PFCs	Subtotal 小计
Scope 3 范围 3						
Methane Generation at landfill due to Disposal of Paper Waste 废纸处理过程中在堆填区产生的甲烷						
Electricity for Processing Fresh Water 淡水处理消耗的电力						
Electricity for Processing Sewage 污水处理消耗的电力						
Others 其他						
Total Scope 3 GHG Emissions 范围 3 碳排放总计						

References
参考文献

[1]About the United Nations Environment Programme [EB/OL]. (2017-07-27) [2022-05-15]. http://www.unep.org/about-us.

[2]Access to a healthy environment, declared a human right by UN rights council | UN News [EB/OL]. (2021-10-08) [2022-05-15]. https://news.un.org/en/story/2021/10/1102582.

[3]Actions for a healthy planet [EB/OL]. [2023-07-27]. https://www.un.org/en/actnow/ten-actions.

[4]ALMOND R E A, GROOTEN M, PETERSEN T. Living Planet Report 2020-Bending the curve of biodiversity loss[R/OL]. WWF, 2020[2022-05-15]. https://wwfin.awsassets.panda.org/downloads/lpr_2020_full_report.pdf.

[5]An Introduction to Climate Change and Human Rights [EB/OL]. [2022-05-15]. https://unccelearn.org/course/view.php?id=136&page=overview&lang=en.

[6]Basics of Climate Change [EB/OL]. [2023-06-24]. https://www.epa.gov/climatechange-science/basics-climate-change.

[7]BEAM Plus [EB/OL]//Hong Kong Green Building Council Limited. [2022-05-15]. https://www.hkgbc.org.hk/eng/beam-plus/introduction/index.jsp.

[8]Beijing 2022 Pre-Games Sustainability Report[R/OL]. Beijing Organising Committee for the 2022 Olympic and Paralympic Winter Games, 2022[2023-06-24]. https://stillmed.olympics.com/media/Documents/Olympic-Games/Beijing-2022/Sustainability/Beijing-2022-Pre-Games-Sustainability-Report.pdf.

[9]California Prepares for Increased Wildfire Risk to Air Quality From Climate Change[EB/OL]//US EPA. (2023-02-22) [2023-06-24]. www.epa.gov/arc-x/california-prepares-increased-wildfire-risk-air-quality-climate-change.

[10]Causes and Effects of Climate Change [EB/OL]. [2022-05-15]. https://www.un.org/en/climatechange/science/causes-effects-climate-change.

[11]Clean Air Plan for Hong Kong 2035[R/OL]. Environment and Ecology Bureau, The Government of the Hong Kong Special Administrative Region, 2021[2022-05-15]. https://www.eeb.gov.hk/sites/default/files/pdf/Clean_Air_Plan_2035_eng.pdf.

[12]Climate Action Fast Facts [EB/OL]//United Nations. [2022-05-15]. https://www.un.org/en/climatechange/science/key-findings.

[13]Climate Change in Hong Kong[EB/OL]//Hong Kong Observatory. (2021-07-27)[2022-05-15]. https://www.hko.gov.hk/en/climate_change/climate_change_hk.htm.

[14]Climate Change: From Learning to Action [EB/OL]//UN CC: Learn. [2022-05-15]. https://unccelearn.org/course/view.php?id=48&page=overview&lang=en.

[15]Climate Effects on Health | CDC[EB/OL]//U.S. Department of Health & Human Services. [2022-05-15]. https://www.cdc.gov/climateandhealth/effects/default.htm.

[16]Council for Sustainable Development [EB/OL]//Environment and Ecology Bureau, The Government of the Hong Kong Special Administrative Region. [2022-05-15]. https://www.enb.gov.hk/en/susdev/council/index.htm.

[17]Each Country's Share of CO_2 Emissions[EB/OL]. [2022-05-15]. https://www.ucsusa.org/resources/each-countrys-share-CO_2-emissions.

[18]Energyland-Greenhouse gases[EB/OL]. [2022-05-15]. https://www.emsd.gov.hk/energyland/en/energy/environment/greenhouse.html.

[19]EPA. Sources of Greenhouse Gas Emissions[EB/OL]//United States Environmental Protection Agency. (2022) [2022-05-15]. https://www.epa.gov/ghgemissions/sources-greenhouse-gas-emissions.

[20]FLATICON. Flaticon, the largest database of free vector icons[EB/OL]//Flaticon. (2010). https://www.flaticon.com/.

[21]Food systems account for over one-third of global greenhouse gas emissions | UN News[EB/OL]//United Nations. (2021-03-09)[2023-06-24]. https://news.un.org/en/story/2021/03/1086822.

References
参考文献

[22]Food Waste Challenge | Environmental Protection Department[EB/OL]. [2022-05-15]. https://www.epd.gov.hk/epd/english/environmentinhk/waste/prob_solutions/food_waste_challenge.html.

[23]Food Waste Index Report 2021[R/OL]. Nairobi 00100, Kenyay: United Nations Environment Programme, 2021[2022-05-15]. https://www.unep.org/resources/report/unep-food-waste-index-report-2021.

[24]Forests and climate change[EB/OL]//IUCN. (2021-07-27)[2022-05-15]. https://www.iucn.org/resources.

[25]Frequently Asked Questions (FAQs)-U.S. Energy Information Administration (EIA)[EB/OL]. [2022-05-15]. https://www.eia.gov/tools/faqs/faq.php.

[26]Global Effects of Mount Pinatubo[EB/OL]. [2023-06-24]. https://earthobservatory.nasa.gov/images/1510/global-effects-of-mount-pinatubo.

[27]Global Monitoring Laboratory-Carbon Cycle Greenhouse Gases[EB/OL]//National Oceanic and Atmospheric Administration. [2022-05-15]. https://gml.noaa.gov/ccgg/trends/.

[28]Greenhouse Effect[EB/OL]//Hong Kong Observatory. (2019-07-27)[2022-05-15]. https://www.hko.gov.hk/en/cis/climchange/grnhse.htm.

[29]GUO J, KUBLI D, SANER P. The economics of climate change: no action not an option[R/OL]. Swiss Re Management Ltd, Swiss Re Institute, 2021[2022-05-15]. https://www.swissre.com/dam/jcr:e73ee7c3-7f83-4c17-a2b8-8ef23a8d3312/swiss-re-institute-expertise-publication-economics-of-climate-change.pdf.

[30]HAGELBERG N. Air pollution and climate change: two sides of the same coin[EB/OL]//UN Environment. (2019-04-23)[2023-06-20]. https://www.unep.org/news-and-stories/story/air-pollution-and-climate-change-two-sides-same-coin.

[31]HOFFMANN C, HOEY M, ZEUMER B. Decarbonization in steel[EB/OL]//McKinsey & Company. (2021-07-27)[2022-05-15]. https://www.mckinsey.com/industries/metals-and-mining/our-insights/decarbonization-challenge-for-steel.

[32]HONG KONG GREEN BUILDING COUNCIL. What is Green Building?[EB/OL]//www.hkgbc.org.hk. (2022)[2022-05-15]. https://www.hkgbc.org.hk/eng/about-us/what-is-green-building/index.jsp.

[33]Hong Kong Roadmap on Popularisation of Electric Vehicles[R/OL]. Environment and Ecology Bureau, The Government of the Hong Kong Special Administrative Region, 2021[2022-05-15]. https://www.evhomecharging.gov.hk/downloads/ev_booklet_en.pdf.

[34]Hong Kong's Climate Action Plan 2050[R/OL]. Environment and Ecology Bureau, The Government of the Hong Kong Special Administrative Region, 2021[2022-05-15]. https://cnsd.gov.hk/wp-content/uploads/pdf/CAP2050_leaflet_en.pdf.

[35]Hong Kong's Climate Action Plan 2030+ report[R/OL]. Environment and Ecology Bureau, The Government of the Hong Kong Special Administrative Region, 2017[2022-05-15]. https://www.hkgbc.org.hk/eng/engagement/file/ClimateActionPlanEng.pdf.

[36]How does climate change affect coral reefs?[EB/OL]//US Department of Commerce. (2015-07-27)[2022-05-15]. https://oceanservice.noaa.gov/facts/coralreef-climate.html.

[37]How Much Do Our Wardrobes Cost to the Environment?[EB/OL]. (2019-09-23)[2022-05-15]. https://www.worldbank.org/en/news/feature/2019/09/23/costo-moda-medio-ambiente.

[38]IPCC . Climate Change 2013: The Physical Science Basis. Contribution of Working Group I to the Fifth Assessment Report of the Intergovernmental Panel on Climate Change[R/OL]//IPCC. Cambridge, United Kingdom and New York, NY, USA: Cambridge University Press, 2013[2022-05-15]. https://www.ipcc.ch/report/ar5/wg1/.

[39]Iron & steel[EB/OL]//IEA. [2022-05-15]. https://www.iea.org/energy-system/industry/steel.

[40]Kingdom of Bhutan Intended Nationally Determined

References
参考文献

Contribution[R/OL]. Royal Government of Bhuta, 2015[2022-05-15]. https://policy.asiapacificenergy.org/sites/default/files/Bhutan-INDC-20150930.pdf.

[41]KUMAR P, SINGH D P. Solar Cycle Variability and Global Climate Change[J/OL]. 2019, 10(4). https://www.omicsonline.org/open-access/solar-cycle-variability-and-global-climate-change.pdf. DOI:10.4172/2157-7617.1000514.

[42]KUMAR P, SINGH D P. Solar Cycle Variability and Global Climate Change[J/OL]. Journal of Earth Science & Climatic Change, 2019, 10(4). https://www.omicsonline.org/open-access/solar-cycle-variability-and-global-climate-change-2157-7617-1000514-109027.html.

[43]LINDSEY R. CLIMATE CHANGE: ATMOSPHERIC Carbon Dioxide[EB/OL](2023-05-12)[2023 - 05-15]. http://www.climate.gov/news-features/understanding-climate/climate-change-atmospheric-carbon-dioxide.

[44]MASSON-DELMOTTE V, ZHAI P, PIRANI A, et al. Climate Change 2021: The Physical Science Basis[R/OL]//IPCC. Cambridge, United Kingdom and New York, NY, USA: Cambridge University Press, 2021[2022-05-15]. https://www.ipcc.ch/report/ar6/wg1/.

[45]MASSON-DELMOTTE V, ZHAI P, PÖRTNER H O, et al. Global Warming of 1.5℃ : Summary for Policymakers[R/OL]//IPCC. Cambridge, UK and New York, NY, USA: Cambridge University Press, 2022[2022-05-15]. https://doi.org/10.1017/9781009157940.001.

[46]MTR CORPORATION LIMITED. MTR Sustainability Report 2021-Sustainable Investment[EB/OL]//www.mtr.com.hk. (2021)[2022-05-15]. https://www.mtr.com.hk/sustainability/en/sustainable-investment.html.

[47]NATIONAL GEOGRAPHIC SOCIETY. Nuclear Energy[EB/OL]//National Geographic Society. National Geographic Society, 2012[2022-05-15]. https://www.nationalgeographic.org/encyclopedia/nuclear-energy/.

[48]NATIONS U. United Nations Summit on Sustainable Development[EB/OL]//United Nations. (2015)[2022-05-15]. https://www.un.org/en/conferences/environment/newyork2015.

[49]Norwegian EV policy[EB/OL]//Norsk elbilforening. [2022-05-15]. https://elbil.no/english/norwegian-ev-policy/#:~:text=The%20Norwegian%20Parliament%20has%20decided.

[50]One Stop Shop | Waste Reduction[EB/OL]. [2022-05-15]. https://www.wastereduction.gov.hk/en-hk/one-stop-shop.

[51]PACHAURI R K, MEYER L A. Climate Change 2014: Synthesis Report[R/OL]. Geneva, Switzerland: IPCC, 2015[2022-05-15]. https://www.ipcc.ch/site/assets/uploads/2018/05/SYR_AR5_FINAL_full_wcover.pdf.

[52]PEARCE R. Mapped: The World's Coal Power Plants in 2020[EB/OL](2020-03-26)[2022-05-15]. https://www.carbonbrief.org/mapped-worlds-coal-power-plants/.

[53]PÖRTNER H O, ROBERTS D C, TIGNOR M, et al. Climate Change 2022: Impacts, Adaptation and Vulnerability[R/OL]//IPCC. Cambridge, United Kingdom and New York, NY, USA: Cambridge University Press, 2022[2022-05-15]. https://www.ipcc.ch/report/ar6/wg2/.

[54]Potential climate changes impact | GRID-Arendal[EB/OL]//www.grida.no. [2022-05-15]. https://www.grida.no/resources/6891.

[55]SCHMIDT T, ALLISON C, ILIFFE M, et al. Transition risk framework: Managing the impacts of the low carbon transition on infrastructure investments[R/OL]. UK: Cambridge Institute for Sustainability Leadership (CISL), 2019[2022-05-15]. https://www.cisl.cam.ac.uk/resources/sustainable-finance-publications/transistion-risk-framework-managing-the-impacts-of-the-low-carbon-transition-on-infrastructure-investments.

[56]SHUKLA P R, SKEA J, CALVO BUENDIA E, et al. Climate Change and Land: an IPCC special report on climate change, desertification, land degradation, sustainable land management, food security, and greenhouse gas fluxes in terrestrial ecosystems[R/OL]. IPCC, 2019[2022-05-15]. https://www.ipcc.ch/

References
参考文献

site/assets/uploads/2019/11/SRCCL-Full-Report-Compiled-191128.pdf.

[57]SOLOMON S, QIN D, MANNING M, et al. Climate Change 2007: The Physical Science Basis[R/OL]//IPCC. Cambridge, United Kingdom and New York, NY, USA: Cambridge University Press, 2007[2022-05-15]. https://www.ipcc.ch/site/assets/uploads/2018/05/ar4_wg1_full_report-1.pdf.

[58]SOUTER D, PLANES S, WICQUART J, et al. Status of Coral Reefs of the World: 2020[J/OL]. 2021[2022-05-15]. https://gcrmn.net/wp-content/uploads/2023/04/GCRMN_Souter_et_al_2021_Status_of_Coral_Reefs_of_the_World_2020_V1.pdf. DOI:https://doi.org/10.59387/wotj9184.

[59]SYMON C. Climate Change: Action, Trends and Implications for Business[R/OL]. University of Cambridge Institute for Sustainability Leadership, 2013[2022-05-15]. https://www.cisl.cam.ac.uk/system/files/documents/Science_Report__Briefing__WEB_EN.pdf.

[60]THOMAS B, FISHWICK M, JOYCE J, et al. A Carbon Footprint for UK Clothing and Opportunities for Savings[R]. Environmental Resources Management Limited (ERM), 2012[2022-05-15].

[61]UNFCCC. The Paris Agreement[EB/OL]//United Nations Framework Convention on Climate Change. United Nations, 2016[2022-05-15]. https://unfccc.int/process-and-meetings/the-paris-agreement/the-paris-agreement.

[62]UNITED NATIONS CLIMATE CHANGE. What is the United Nations Framework Convention on Climate Change? | UNFCCC[EB/OL]//Unfccc.int. (2012)[2022-05-15]. https://unfccc.int/process-and-meetings/the-convention/what-is-the-united-nations-framework-convention-on-climate-change.

[63]UNITED NATIONS. Net Zero Coalition[EB/OL]//United Nations. (2022)[2022-05-15]. https://www.un.org/en/climatechange/net-zero-coalition.

[64]UNITED NATIONS. Transforming our world: The 2030 agenda for sustainable development[EB/OL]//United Nations. (2015)[2022-05-15]. https://sdgs.un.org/2030agenda.

[65]UNITED NATIONS. United Nations Conference on the Environment, Stockholm 1972[EB/OL]//United Nations. (2022)[2022-05-15]. https://www.un.org/en/conferences/environment/stockholm1972.

[66]UNITED NATIONS. What is the Kyoto Protocol? [EB/OL]//UNFCCC. UNFCCC, 2019[2022-05-15]. https://unfccc.int/kyoto_protocol.

[67]UNSPLASH. Beautiful Free Images & Pictures[EB/OL]//Unsplash.com. Unsplash, 2022. https://unsplash.com/.

[68]US EPA. Understanding Global Warming Potentials | US EPA[EB/OL]//US EPA. (2023-04-18)[2022-05-15]. https://www.epa.gov/ghgemissions/understanding-global-warming-potentials.

[69]Waste Blueprint for Hong Kong 2035[R/OL]//https://www.eeb.gov.hk/sites/default/files/pdf/waste_blueprint_2035_eng.pdf. Environment and Ecology Bureau, The Government of the Hong Kong Special Administrative Region, 2021[2021-05-15]. https://www.eeb.gov.hk/sites/default/files/pdf/waste_blueprint_2035_eng.pdf.

[70]WORLD BUSINESS COUNCIL FOR SUSTAINABLE DEVELOPMENT, WORLD RESOURCES INSTITUTE. The greenhouse gas protocol : a corporate accounting and reporting standard.[M]. Geneva, Switzerland: World Business Council For Sustainable Development ; Washington, Dc, 2004.

[71] 习近平在第七十六届联合国大会一般性辩论上的讲话（全文）[EB/OL]. 中国政府网，新华社，(2021-09-22)[2023-10-18]. https://www.gov.cn/xinwen/2021-09/22/content_5638597.htm.

[72] 中华人民共和国国民经济和社会发展第十四个五年规划和 2035 年远景目标纲要 [EB/OL]. 中国政府网，新华社，(2021-03-13)[2023-10-18]. https://www.gov.cn/xinwen/2021-03/13/content_5592681.htm.

[73] 中国十大最节能低碳建筑TOP10[EB/OL]//www.gbwindows.